EVOLUTION AS NATURAL HISTORY

**Recent Titles in
Human Evolution, Behavior, and Intelligence**

Why Race Matters: Race Differences and What They Mean
Michael Levin

The Evolution of Love
Ada Lampert

The G Factor: The Science of Mental Ability
Arthur R. Jensen

Sex Linkage of Intelligence: The X-Factor
Robert Lehrke

Separation and Its Discontents: Toward an Evolutionary Theory of Anti-Semitism
Kevin MacDonald

The Biological Origins of Art
Nancy E. Aiken

The Culture of Critique: An Evolutionary Analysis of Jewish Involvement in Twentieth-Century Intellectual and Political Movements
Kevin MacDonald

Relating in Psychotherapy: The Application of a New Theory
John Birtchnell

The Evolution of the Psyche
D. H. Rosen and M. C. Luebbert, editors

Mind and Variability: Mental Darwinism, Memory, and Self
Patrick McNamara

The Darwinian Heritage and Sociobiology
Johan M. G. van der Dennen, David Smillie, and Daniel R. Wilson

The Culture of Sexism
Ignacio L. Götz

EVOLUTION AS NATURAL HISTORY
A Philosophical Analysis

Wim J. van der Steen

Human Evolution, Behavior, and Intelligence
Seymour W. Itzkoff, Series Editor

Westport, Connecticut
London

Library of Congress Cataloging-in-Publication Data

van der Steen, Wim J., 1940–
    Evolution as natural history: a philosophical analysis / Wim J. van der Steen.
        p. cm. — (Human evolution, behavior, and intelligence, ISSN 1063–2158)
    Includes bibliographical references and index.
    ISBN 0–275–96870–7 (alk. paper)
    1. Evolution (Biology)—Philosophy.   I. Title.   II. Series.
QH360.5.S74 2000
576.8′01—dc21   99–052983

British Library Cataloguing-in-Publication Data is available.

Copyright © 2000 by Wim J. van der Steen

All rights reserved. No portion of this book may be
reproduced, by any process or technique, without the
express written consent of the publisher.

Library of Congress Catalog Card Number: 99–052983
ISBN: 0–275–96870–7 (alk. paper)
ISSN: 1063–2158

First published in 2000

Praeger Publishers, 88 Post Road West, Westport, CT 06881
An imprint of Greenwood Publishing Group, Inc.
www.praeger.com

Printed in the United States of America

The paper used in this book complies with the
Permanent Paper Standard issued by the National
Information Standards Organization (Z39.48–1984).

10 9 8 7 6 5 4 3 2 1

Acknowledgments

I have used, in revised form, materials from my previously published articles, with kind permission from Kluwer Academic Publishers [Altruism and egoism in ethics: Dispensing with spurious generality, *Journal of Value Inquiry* 29, 31–44, 1995; Evolution and altruism, *Journal of Value Inquiry* 33, 11–29, 1999; Methodological problems in evolutionary biology X: Natural selection without selective agents, *Acta Biotheoretica* 46, 99–107, 1998; Bias in behavior genetics: An ecological perspective, *Acta Biotheoretica* 46, 369–377, 1998; Methodological problems in evolutionary biology XI: Optimal foraging theory revisited, *Acta Biotheoretica* 46, 321–336, 1998; Methodological problems in evolutionary biology XII: Against evolutionary ethics, *Acta Biotheoretica* 47, 47–57, 1999], from *History and Philosophy of the Life Sciences* [New ways to look at fitness, vol. 16, 479–492, 1994], from Blackwell Science [Dissolving disputes over genic selectionism, *Journal of Evolutionary Biology* 12, 184–187, 1999, with H. A. van den Berg], and from the University of Chicago Press [Screening-off and natural selection, *Philosophy of Science* 63, 113–119, 1996; © copyright 1996 by the Philosophy of Science Association. All rights reserved].

Gratefully I thank the following colleagues, who by constructive comments helped me improve the text: David Hull, Peter Kirschenmann, Bas Kooijman, Theo Kuipers, Elliott Sober, and Cor Zonneveld.

Contents

1	**Introduction**		1
2	**Problems with Fitness**		7
	2.1	Introduction	7
	2.2	Preliminaries: The Generality of Biological Theories	8
	2.3	Fitness as Reproductive Survival: No Tautologies	11
	2.4	Fitness as a Supervenient Concept	12
	2.5	Supervenience Made Concrete	13
	2.6	The Propensity Interpretation: Mills and Beatty	16
	2.7	The Propensity Interpretation: Sober	17
	2.8	Some Recent Developments	18
	2.9	Conclusions	22
3	**Adaptationism**		25
	3.1	Introduction	25
	3.2	Analyzing Riddles	26
	3.3	Selection without Selective Agents	28
	3.4	Conclusions	36
4	**The Chimera of Optimality**		39
	4.1	Introduction	39

	4.2	Generality, Universality, and Testability	40
	4.3	Constraints and Free Will	42
	4.4	Hard and Soft Constraints	45
	4.5	Controversies over Constraints	49
	4.6	Conclusions	54
5	**The Units of Selection**		55
	5.1	Introduction	55
	5.2	Genic Selectionism: The Issue of Representation	56
	5.3	The Puzzle of Screening-off	61
	5.4	Group Selection and Species Selection	65
	5.5	Conclusions	71
6	**Evolution and Altruism**		73
	6.1	Introduction	73
	6.2	Egoism and Altruism: A Scheme for Conceptual Analysis	74
	6.3	Egoism versus Altruism: Samples from Ethics	76
	6.4	The Relevance of Empirical Issues	80
	6.5	Explaining Altruism in Animals	82
	6.6	The Paradigm of Evolutionary Psychology	85
	6.7	The Paradox of Altruism: Sesardic's View	89
	6.8	Dissolving the Paradox of Altruism	92
	6.9	Conclusions	96
7	**Evolution and Culture**		99
	7.1	Introduction	99
	7.2	Natural and Cultural Selection	100
	7.3	Against Overarching Theories of Culture	103
	7.4	Evolutionary Psychologists on Culture	105
	7.5	Conclusions	110
8	**Against Evolutionary Ethics**		111
	8.1	Introduction	111
	8.2	Demise of the Naturalistic Fallacy?	112
	8.3	No Foundations for Ethics?	121
	8.4	The Explanatory Relevance of Evolutionary Biology	125
	8.5	Evolutionary Thinking in Normative Settings	127
	8.6	Conclusions	128

9	**Evolution and Knowledge**		131
	9.1	Introduction	131
	9.2	The Evolution of Cognition	132
	9.3	The Evolution of Scientific Theories	136
	9.4	Toward a Broader Perspective	139
	9.5	Conclusions	146
10	**Diseases in an Evolutionary Perspective**		149
	10.1	Introduction	149
	10.2	The Need for Darwinian Medicine	150
	10.3	Natural Selection and Psychopathology	152
	10.4	The Environment in Aggression and Psychopathology	157
	10.5	Conclusions	160
11	**Conclusions**		161

References   165
Index   179

1

Introduction

All species, our own included, have been shaped by evolutionary transformations. Almost all researchers accept that nowadays. The theory of explaining evolution is a different matter. Controversies over the theory, or particular variants of it, are common.

To assess the theory, we need criteria for theoryhood. From a traditional point of view, scientific theories consist of interconnected, general, universal laws of nature which are well confirmed. It is commonly taken for granted that such laws exist in physics and that biology compares poorly with physics in this respect. But this must be qualified. Cartwright (1983) has argued convincingly that the general laws of physics do not describe the real world. For realistic descriptions, we need phenomenological laws at low levels of generality. Such laws may be universal and well confirmed, but they apply to restricted domains. They only follow from general laws together with additional assumptions and emendations.

This characterization also applies to some parts of biology. For example, biological systems should satisfy general laws of physics concerning mass and energy conservation. If we supplement the laws with information about flows of mass and energy in living systems, we get the equivalent of phenomenological laws in physics for biology. The theory of dynamic energy budgets in biological systems developed by Kooijman (1993) illustrates this.

Most areas of biology do not even have universally valid laws at low levels of generality. Lawton (1999) argues that generalizations of ecology are not universally valid. Brandon (1996) holds that biology characteristically yields "contingent generalizations," which are not laws, but he emphasizes that such generalizations have explanatory value. Considering laws in ecology, Weber

(1999) is less pessimistic, but he only elaborates one possible law—the principle of competitive exclusion, which is controversial.

My concern is with the status of evolutionary theory. Considering the feasibility of laws in biology, we should put this theory high on our agenda because evolutionary thinking pervades almost all areas of biology. I aim to locate proper levels of generality for theorizing in evolutionary biology.

Beatty argues that evolutionary theory explains why few, if any, laws exist in biology. The following passage captures his "evolutionary contingency thesis."

Under what circumstances could we reasonably expect a single theory to suffice for a domain of biological phenomena? To expect a single mechanism underlying an entire domain of biological phenomena, we would have to assume that one mechanism evolved in a common ancestor of all the taxa covered by the domain, and that the mechanism has been maintained in each of those taxa ever since, and/or we would have to assume that the very same mechanism arose independently and has been maintained in all the taxa covered by the domain. In other words, we would have to assume *extreme phylogenetic conservatism*, and/or extremely strong and remarkably similar selection pressures resulting in *extreme parallel evolution* [Beatty 1996: S436].

In brief, evolution produces diversity not easily covered by general laws.

Sober (1996) criticizes Beatty's evolutionary contingency thesis. Beatty's point is that contingent initial conditions existing at some point in evolutionary time determine phenomena occurring at a later point in time. Sober notes that we can accommodate these conditions in statements with the form "If such-and-such initial conditions exist at some point in time, then such-and-such phenomena will occur at a later point in time." Such statements may be regarded as laws. Sober may well be right, but he does not provide examples of such laws. For the time being, we have to accept that laws are scarce in biology.

Process structuralists, most notably Goodwin (1994), envisage other possibilities for developing laws in biology. They oppose modern evolutionary biology on the ground that it does not provide laws. The emphasis in evolutionary biology is on historical changes in genetic factors, which can only explain differences between organisms in particular features. It does not explain the order underlying the diversity of organic forms. Goodwin proposes that we should aim at field equations that explain the generation of form during development. Such equations determine the morphospace occupied by generic forms of organisms. Genes and environmental factors have only a marginal role in that they set the values of parameters in the equations. As yet, the structuralists have not elaborated the body of laws they envisage. Their ideal is indeed problematic because the generic forms that exist in nature presumably represent a tiny subset of the forms that would be allowed by natural selection. The forms existing now are the outcome of historically contingent processes (Griffiths 1996a).

Yet sweeping claims are common in evolutionary thinking. To begin with, some hold that evolutionary biology is built around the principle of natural

Introduction 3

selection as a well-confirmed law. Furthermore, we hear that features of organisms are typically an adaptive result of natural selection, that organisms typically perform optimal behaviors, that natural selection makes genuine altruism impossible, and that evolutionary theory is a proper foundation for ethics, epistemology, and other disciplines.

I argue that such generalities must be distrusted, since evolutionary theory is best reconstructed as natural history. I use the term "natural history" for bodies of theory at low levels of generality which often contain nonuniversal claims. I analyze diverse themes of evolutionary biology to uncover natural history.

This book consists of two parts. In chapters 2–5, I stay with evolutionary theory proper. I analyze central concepts of the theory in defense of the thesis that evolutionary thinking is mostly natural history. In chapters 6–10, I consider extensions of evolutionary biology toward other disciplines, to chart what it has to say about our own species.

Chapter 2 deals with the thesis that evolutionary biology is problematic because it relies on a tautological principle of natural selection. It is easy to rebut this charge, because a general, empirical principle of natural selection does not exist in texts of evolutionary biology. We are dealing with an odd fabrication. Central in the debates about tautologies are problems with the concept of fitness. I argue that we should distinguish between two kinds of fitness concept: fitness as expected reproductive success, and engineering fitness. Fitness as reproductive success does generate tautologies if it is invoked to explain reproductive success, but this is a silly sort of explanation. Fitness in this sense rather explains, together with other factors, the fate of types in populations. Engineering fitness, in turn, explains reproductive success. More accurately, features of organisms representing engineering fitness, different ones in different cases, explain the reproductive success of organisms in particular environments. Engineering fitness as such is an empty placeholder term without explanatory force. Substantive explanations thus represent natural history.

As argued in chapter 3, "natural selection" is likewise a placeholder concept. It covers all sorts of selective agents. The mere thesis that particular evolutionary processes represent natural selection is uninformative without natural history details about selective agents. These details supplement details concerning features of organisms representing engineering fitness. We should beware thinking that selective agents *always* play a role in selection processes. Many researchers assume, implicitly if not explicitly, that selection is impossible without selective agents. This thesis is false. Chapter 3 also considers adaptationism, the thesis that features of organisms are typically the adaptive result of natural selection. I argue that disputes over adaptationism are misguided due to conceptual problems with adaptation.

Links exist between adaptationism and the thesis of optimality, the subject of chapter 4. I focus on the theory of optimal foraging, OFT. If this theory is construed as the thesis that most if not all foraging behaviors of animals are optimal, it is meaningless. Optimality in general is a worthless notion. Lizards do not forage on insects up in the air—they can't fly. If this kind of thing would point to nonoptimality, optimal behaviors of animals would not exist at all. We apparently have to consider optimality relative to particular constraints. Oddly, all behaviors become optimal under an inclusive definition of constraint. Hence, we are left with specific models showing that particular behaviors are optimal relative to a particular set of constraints, while they would be nonoptimal relative to a different set of constraints. Thus, we again get natural history rather than general theory. This is not to deny, though, that OFT has a general qualitative core saying that grossly maladaptive foraging behaviors are rare or nonexistent. The core of OFT is well confirmed, but it is not very informative.

Models of population genetics in evolutionary biology come closest to the ideal of generality. These models mostly describe evolutionary processes as changes in gene frequencies. As such, they do not generate full-fledged explanations. But some researchers have argued that the genic perspective is all-important. The thesis that evolution is best conceptualized as genetic change is known as genic selectionism. In chapter 5, I argue that disputes over genic selectionism are futile. Equations of population genetics describing changes in gene frequencies as mere equations are uninformative. If their intended empirical content is spelled out, reference to causal processes involving individual organisms becomes unavoidable. Genic selectionists forget about an essential context of the equations, which typically takes the form of natural history. Opponents of genic selectionists have argued that individuals rather than genes are the prime target of selection. This view is as suspect conceptually as is genic selectionism. Effects on individuals and effects on genes cannot be disentangled. Genes and individuals are equally important in evolutionary change. Groups and species may also play causal roles in selection processes. Unlike the disputes over genes versus individuals, disputes about these higher-level entities have substance, because higher levels need not play a role in all selection processes. But precise ways to conceptualize group selection and species selection are arbitrary to some extent. We should distinguish between different causal processes involving groups and species.

Chapter 6 considers the thesis that altruism is at odds with evolutionary biology. This claim, like the claims considered in previous chapters, is problematic for conceptual rather than factual reasons. To begin with, notions of altruism in evolutionary biology and notions of altruism in ethics and in daily life have different meanings. The alleged impossibility of evolutionary altruism, therefore, does not imply that altruism in mundane senses is impossible. Furthermore, the emphasis in evolutionary biology is overmuch on selfishness. Explanations of apparent altruism mostly reduce it to selfishness, while overlooking that many forms of cooperation are evolutionarily feasible. Apart from this, the thesis that all human behaviors are egoistic in a mundane sense—psychological egoism—is obscure, and likewise for the thesis that this is as it should be—ethical egoism. Egoism promotes advantages for oneself at costs for others. If the notion of advantage is given a broad interpretation, psychological egoism becomes a

Introduction

boring tautology, so that ethical egoism is pointless. To make sense of psychological egoism and ethical egoism, we must specify which things are to count as advantages. That can be done in different ways. We have to distinguish between different forms of egoism and altruism. Thus, empirical claims about egoism in general and altruism in general are not meaningful. The theme calls for natural history rather than general theory.

Human beings are special in that they have culture. In principle, evolutionary biology can explain some aspects of culture, but not all aspects. The explanation of particular aspects of culture should take the form of natural history including historical data. In chapter 7, I review relations between evolutionary change and cultural change, which come in different kinds. Some theories model cultural change as a selection process that is analogous to natural selection. I argue that this type of theory is problematic because the category of cultural change is heterogeneous, much more so than the category of biological evolution. Hence, the thesis that cultural change is a process of selection is uninformative.

Evolutionary biology consists of empirical claims that, as such, do not allow the derivation of normative claims. Such derivations would constitute a naturalistic fallacy. Yet, some researchers maintain that evolutionary foundations are possible for normative ethics and normative epistemology. I criticize their claims in chapters 8 and 9, respectively. Also, I comment on more modest claims invoking an evolutionary explanation of the origin of norms in ethics and epistemology. Considering ethics, I argue that the lack of an evolutionary foundation does not make evolutionary approaches irrelevant. Knowledge of evolutionary processes is vital for health care and environmental policy. It is unfortunate that evolutionary ethicists disregard these topics.

In chapter 10, I consider implications of evolutionary thinking for medicine. The recent emergence of Darwinian medicine should be applauded. But ideas promoted by researchers in evolutionary psychiatry are problematic. They characteristically assume that psychiatric disorders are so common that they must have been adaptive in the recent past. Underlying this idea is the assumption that specific genetic factors play an important role in the etiology of the disorders. I criticize this assumption, and also the corollary that the disorders must have been adaptive.

The chapters in the book combine to show that conceptual analysis should play a major role in the assessment of evolutionary thinking. Overall, my own analyses uncover much natural history in evolutionary biology.

2

Problems with Fitness

2.1 INTRODUCTION

Barring exceptions such as creationists, few researchers quarrel with the thesis that life as we know it is the result of evolutionary processes spanning aeons. The evidence confirming the phenomenon of evolution is diverse and overwhelming. But the theory explaining evolution is more controversial.

To assess the theory, we need criteria for theoryhood. From a classical point of view, theories consist of well-confirmed laws of nature; laws themselves are universal, highly general empirical theses supported by incontrovertible evidence. Throughout this book, I argue that evolutionary theory is at variance with this view. Biologists have not been able to formulate any law of evolution. Instead, evolutionary biology comprises a rich variety of context-dependent, nongeneral empirical claims best characterized as "natural history." True, we should endorse some qualitative general claims about evolution—for example, the claim that natural selection has been an important force in many evolutionary processes. But such claims lack the preciseness that should characterize laws of nature.

Evolutionary biologists nonetheless aim, of course, to attain the highest level of generality achievable. I argue that they often overshoot the mark and end up with spurious generality. Likewise for philosophers of biology concerned with evolution.

In the present chapter, I consider the claim that evolutionary biology has at its core a general principle of natural selection that qualifies as a law. The catchphrase "The fittest survive" has often served to represent such a law. It has generated heated debates spanning decades. Critics of evolutionary biology have argued as follows: "The fittest survive" is at the core of evolutionary theory. It is a tautology since fitness is defined by way of survival. The core of theories should not contain tautologies. Hence evolutionary theory is no good.

I argue that the tautology charge distorts evolutionary biology. We can easily rebut it once we realize that the notion of survival is ambiguous. Fitness in one sense has to do with reproductive success. It is obvious that fitness in this sense cannot explain survival as reproductive success. Instead, the proper *explanandum* should be survival representing the persistence of types in populations. Fit-ness and other factors such as mutation and migration properly explain changes in populations. The explanations are in the domain of population genetics, a sub-discipline of evolutionary biology.

To get at full-fledged evolutionary explanations, we should also explain differences in fitness as reproductive success; we can do that by supplementing population genetics with ecology. Confusingly, this type of explanation often resorts to a different fitness concept: engineering fitness or fitness as design. Organisms that are reproductively successful have a design that ensures reproduction. In brief, differences in engineering fitness, together with environmental factors, explain differences in fitness as reproductive success, and differences in reproductive success, together with other factors, explain differential survival of types in populations.

However, the explanatory value of engineering fitness is limited, because "engineering fitness" is an elusive notion The features representing good design are different in different species. Hence, we cannot define "engineering fitness" by the specification of features representing designs. Indeed we cannot define it at all. "Engineering fitness" is best regarded as a placeholder notion which represents different features in different contexts. This is one of the reasons why evolutionary explanations, if spelled out, cannot be very general.

Philosophers of biology have elaborated yet another concept of fitness, the so-called propensity concept. This concept resembles the engineering concept, but it is defined as design *for reproductive survival*, implicitly if not explicitly. I argue that the propensity concept does generate tautologies and circular reasoning. This is odd, because the aim to overcome the tautology problem has been a motive for introducing the concept.

Fitness concepts, upon a proper explication of evolutionary thinking, do not generate tautologies and circles. A proper explication does reveal that evolutionary biology is mostly natural history.

2.2 PRELIMINARIES: THE GENERALITY OF BIOLOGICAL THEORIES

The assertion that the fittest survive has been linked up with Darwin's work, but wrongly so. Darwin introduced the phrase "survival of the fittest" in the fourth edition of *On the Origin of Species*, on Spencer's advice. In the present century, "survival of the fittest," a mere catchphrase, was turned into a thesis, "The fittest survive," which has generated much confusion.

In discussions concerning fitness, the tautology problem is an important issue. Critics of evolutionary theory have argued that the core of the theory has no empirical content (is a tautology) due to conceptual links between "fitness" and "survival." They have been answered in various ways, but no consensus has been reached.

"Fitness" as a general concept is designed to play a crucial role in general theories of biology. The search for generality is an old ideal of scientists and philosophers alike. Elsewhere I have argued that the ideal is overemphasized; many methodological criteria in addition to generality play a role in the assessment of theories (van der Steen 1993). Generality is important, but it may be at odds with other criteria that are more important in some contexts. Biology indeed contains more "natural history" than highly general theory (van der Steen and Kamminga 1991).

Most philosophers would use the term "theory" for highly general bodies of empirical laws of nature. Biology has few if any theories in this sense. However, the so-called semantic view in philosophy looks at theories in a different way. Theories in the semantic sense are construed as models of ideal systems without empirical content. The empirical content of scientific work is instead located in theoretical hypotheses stating that particular empirical systems are instantiations of ideal systems.

The following example explains how this works out for biology. Patterns of population growth are a mixed lot. Under some conditions, populations grow exponentially. Under different conditions we find, for example, logistic growth. Could we elaborate theories that describe population growth by generally valid empirical laws? Consider exponential growth. Suppose that a population satisfies the following conditions: (i) that birth rate and death rate (the number of births and the number of deaths per individual per time unit) are constant; and (ii) that other factors such as migration do not affect the number of individuals in the population. These assumptions imply that population growth satisfies the differential equation $dN/dt = a \cdot N(t)$, in which $N(t)$ represents the number of individuals at time t. The solution of the equation is $N(t) = c \cdot e^{at}$. This is an exponential equation. Should the equation be regarded as a law of nature? That sounds implausible. If the equation is taken to apply to all populations, it is not a law, because it is manifestly false. If it is taken to apply to some populations only, it is not a law, for lack of generality. However, we are still left with the option to conditionalize the equation: all populations that satisfy conditions (i) and (ii) grow in accordance with the equation $N(t) = c \cdot e^{at}$. That is a true general thesis. But this thesis does not have empirical content, since growth in accordance with the equations is simply a logical consequence of the two conditions being satisfied. So it is not a law either.

Yet, from the semantic viewpoint, we may regard the equation, in the original form or in the conditionalized form, as a law that describes an ideal system. I have no quarrel with this, but I would regard it as a poor strategy to arrive at laws and theories in biology. The search for laws in the original sense is a search for true, general empirical claims. The decision to use the term "law" for non-empirical claims would not help us with this search. We would still want to get at true, general *empirical* claims. Under the semantic terminology, our aim should become to find proper theoretical hypotheses, the analog of laws in the old sense (Sloep and van der Steen 1987; van der Steen 1993).

Considering population growth, I do think that laws in the old sense can be formulated. I would conjecture, for example, that populations will grow exponentially in the absence of intraspecific and interspecific competition, provided that the environment is constant, migration does not occur, and resources are not in short supply. Under different conditions, we will get different patterns of growth. The patterns are described by mathematical models. The models as such have no empirical content, but they generate empirical claims if expanded with proper conditions of applicability.

A similar situation exists in evolutionary biology. Here we are dealing with evolution in the sense of changes in the composition of populations. Population genetics offers us many models describing such changes, each with its own conditions of applicability.

Population dynamics does not provide an overarching law of population growth. Instead, we must be content with a body of relatively specific "laws." Likewise for population genetics. It does not provide an overarching "law of evolution." Yet, disputes over the tautology problem suggest that the merits of evolutionary biology depend on the existence of such a law, a general principle of natural selection explicating the catchphrase "The fittest survive."

Kitcher (1982: 57–58) has remarked that the expression "natural selection" occurs many times in texts of evolutionary biology, but no general principle of natural selection is ever formulated. This suggests that philosophical disputes about such a principle distort evolutionary biology. The disputes should dissolve under a proper reconstruction of evolutionary biology. The recent adoption of the semantic view by some philosophers of biology (Beatty 1980; Lloyd 1988; Thompson 1989) hampers the dissolution of the disputes, because it detracts from empirical issues. Originally, the issue was whether "The fittest survive" has empirical content. Now we have to face the additional issue of whether fundamental principles (laws, theories) need to have empirical content at all. As indicated, the semantic view would not solve any problem by the stipulation that the principle of natural selection is a law without empirical content. This stipulation amounts to a terminological decision concerning the meaning of "law" which displaces problems with laws to the domain of theoretical hypotheses.

In the next section, I argue that existing disputes have their point of departure in interpretations of "The fittest survive" that do distort evolutionary biology. The tautology problem is a fabrication of philosophers that hampers understanding of evolutionary biology. Subsequently, I analyze reconstructions of evolutionary theory by philosophers of biology aiming to solve problems with "fitness." I argue that these reconstructions create problems where none existed.

2.3 FITNESS AS REPRODUCTIVE SURVIVAL: NO TAUTOLOGIES

In evolutionary biology, especially population genetics, "fitness" in one important sense stands for expected reproductive survival (expected numbers of descendants—survival for short). At first sight this implies that "The fittest survive" is a tautology since fitness is simply survival; thus fitness could not explain survival. This is the original version of the tautology problem, which has generated a huge amount of confusing philosophical literature. I disregard this literature, because the disputants share erroneous views concerning proper meanings of survival. They typically assume that evolutionary biologists wish to explain the reproductive survival of individuals by appealing to the fitness of individuals. But no evolutionary biologist would wish to do that. A cursory inspection of any textbook in population genetics suffices to show that the *explanandum* is the survival of types in populations. The following example indicates that the tautology problem dissolves under a proper interpretation of "survival."

Consider a population with two asexually reproducing types of organism, A and B, in which A consistently has more descendants than B. (I disregard sexuality for ease of exposition. Also disregarded is the role of the environment, which I assume to be static.) If limits to the growth of the population exist, it will ultimately consist of A organisms. Population geneticists describing this would say that A has a greater fitness than B (meaning that A is better at reproductive survival, that it has more descendants, than B), and that this explains the survival of A in the population.

It is easy to rebut the tautology charge, since "survival" has two meanings. Survival in the sense of reproduction of *organisms*, or types of organisms, should be distinguished from survival in the sense of persistence of types in *populations*. In the example, survival in the first sense is one of the factors that explain survival in the second sense, other factors being mutation, migration, and the like. So "The fittest survive," on the most sensible interpretation, is an empirical statement that is true if certain *ceteris paribus* conditions (concerning absence of mutation, migration, frequency-dependent selection, etc.) are satisfied.

This interpretation should not be taken to mean that "The fittest survive," as a principle of natural selection, has the status of an overarching law of evolution. Populations seldom satisfy the *ceteris paribus* conditions. To describe evolutionary processes, we need all sorts of equations describing how fitness, together with other factors, affects populations. At best, we may endorse a qualitative principle of natural selection which says that heritable fitness differences play an important role in most evolutionary processes.

It is convenient to have some name tags for the concept of fitness used here, and for explanations that it provides. Since the context is that of population genetics, I shall use the prefix "pg". Hence we have *pg-fitness*, which figures in *pg-explanations* of population change. Pg-explanations attribute particular changes in the composition of populations to particular differences in pg-fitness and other relevant factors such as mutation and migration.

Pg-fitness should not be defined as *actual* survival. Two organisms may be qualitatively identical and yet by chance have different numbers of descendants. If that is the case, we will still want to say that they have the same pg-fitness. Therefore, fitness must be construed as probability of survival. The tautology problem is not affected by this.

2.4 FITNESS AS A SUPERVENIENT CONCEPT

Pg-explanations represent but one *stage* of evolutionary explanation. Differences in survival (pg-fitness) help to explain changes in populations. The differences themselves can also be explained by reference to features of organisms (together with features of the environment) that are causally responsible for reproductive survival. Hence we should distinguish a second stage of explanation, which I call *e-explanation*, where the "e" stands for "ecological."

Features responsible for survival can also be covered by a fitness concept, one that should not be confused with pg-fitness. I use the term "e-fitness" for this concept. Biologists presumably have e-fitness in mind when they use the term "engineering fitness" (for explications, see Burian 1983).

High *engineering fitness* (*e-fitness*) connotes good design. Now the concept of design, without further specification, is hardly meaningful. Things are not designed *simpliciter*, they are "designed" for a "purpose." What could the purpose be in the case of engineering fitness?

A bad answer would be that engineering fitness consists in design *for survival*. Once we give this answer, we are threatened by tautologies and circular explanations. This is illustrated by the following scheme for e-explanation.

Suppose we want to explain why organisms of a certain type A have more descendants (have a higher pg-fitness) than organisms of type B (in some environment). We could do that by attributing a higher e-fitness to A than to B. The explanation can be cast into the following argument form:

Premise 1: For all x and y, if x has a higher e-fitness than y, then x has more descendants than y.

Premise 2: A has a higher e-fitness than B.

Conclusion: A has more descendants than B.

In this scheme, x and y are variables for types of organisms, and A and B are constants representing particular types.

At first sight, this is an acceptable argument scheme. However, it becomes suspect if e-fitness is conceptually tied to survival. The first premise then means: "For all x and y, if x and y have features that cause x to have more descendants than y, then x has more descendants than y." This is a tautology. For similar reasons, the second premise is suspect. After unpacking, it appears to contain the conclusion. So the explanatory argument is circular. Therefore, design had better be interpreted as design for particular activities, or generic ones such as mating, avoiding predators, and so forth. This interpretation will take different forms in different contexts.

The concept of e-fitness becomes elusive if we avoid the conceptual link with survival. A general *definition* of the concept cannot refer to particular or generic features of organisms, because organisms are designed in different ways. In polar bears, thick fur is an important feature. Needless to say, fur is not relevant in the case of bacteria. Even generic features are not shared by all organisms.

Therefore, we can but construe e-fitness as an abstract concept that is instantiated by different features in different contexts. In philosophical terms, this means that e-fitness is a *supervenient* concept. E-fitness supervenes on manifest features of organisms (and the environment). If two organisms have the same manifest features, and their environments are similar, they should have the same e-fitness. But a particular value of e-fitness can be realized by numerous sets of manifest features. Hence, e-fitness cannot be defined by reference to such features. Indeed, we may need to leave it undefined.

I use the term "placeholder concept" for concepts, such as e-fitness, that take on different meanings by context-dependent specifications. E-fitness is a placeholder for manifest features of organisms, different ones in different contexts. Later chapters will show that e-fitness is by no means the only placeholder concept in evolutionary biology. Analyses of such concepts will indicate that much seemingly general theory needs to be transformed into natural history.

My reconstruction of the concept of e-fitness suppresses the relational component connoted by the concept of design *for* something. Relations (e.g. "for") are best covered by empirical relationships, not conceptual ones. Thus manifest features of organisms covered by the concept of e-fitness can be regarded as *causes*, direct or indirect ones, of survival. Their being causes would be an empirical matter. The analyses in the next section chart the consequences of this reconstruction for explanations.

2.5 SUPERVENIENCE MADE CONCRETE

Suppose that the classical explanation of industrial melanism in the moth *Biston betularia* is adequate. In polluted areas, the trunks of trees on which the moths rest are dark. As a result of this, light morphs of moths are replaced by darker ones, which are better protected against predation.

Important elements of the classical explanation are covered by a reconstruction that distinguishes two "stages." (For the sake of exposition I am suppressing the role of the environment.) First we can pg-explain changes in moth frequencies by noting that dark moths have a higher pg-fitness than light ones. Next, we can e-explain the difference in pg-fitness by showing that it results from differences in morph color and predation (in particular environments).

It is easy to elaborate e-explanations in specific terms (e.g. morph color, predation) that are not plagued by tautologies and circles. However, we may want to have explanations of a more general kind. Would the concept of e-fitness allow us to elaborate more general explanations? I doubt this for the following reasons.

The argument scheme introduced in section 2.4 would cover the pattern of a general e-explanation. The phenomenon to be explained is a particular type A of organism having more descendants (having a higher pg-fitness) than another type B (A and B representing light and dark morphs of *Biston betularia*). We can derive a statement expressing this from a general "law" and a more specific premise. The law would say that, for all x and y, if x has a higher e-fitness than y, then x has more descendants than y. If we add the premise that A has a higher e-fitness than B, we can infer that A has more descendants than B.

This e-explanation is couched in general terms, but if it is unpacked we are forced to make it more specific. The premise that A has a higher e-fitness than B means that *there are* features of A and B that explain differences in numbers of descendants. The presence of an existential component in the premise points to the absence of essential information that should play a role in a full-fledged explanation. The general "law" also contains an existential component with this status. In fact, we are dealing not with a general law but with a scheme that covers a heterogeneous set of causally relevant factors that may influence numbers of descendants. In full-fledged e-explanations, the "law" must be replaced by more specific statements concerning particular causal factors. E-explanations are not general in a substantive way. They are "general" only in the sense that different e-explanations share a common pattern.

At first sight, a statement to the effect that some organism has a high e-fitness is a *statement about a higher-level property* of the organism. If the statement is unpacked, it turns out to be a *higher-level statement about properties*. It says that *there are* properties (not mentioned in the statement) that affect the survival of the organism. The statement does not belong to the domain of substantive empirical theory. It is, rather, an element of metatheory (for similar characterizations, see Darden and Cain 1989; Kitcher 1989).

With respect to the causes of survival, it is desirable to redirect the search for generality. For example, in the explanation of industrial melanism we might appeal to general theories concerning camouflage and predation. This would lead to generalizations over relatively specific domains of ecology.

Pg-fitness and e-fitness do not exhaust extant explications of the notion of fitness. In the rest of this chapter I focus on a third option, the so-called propensity interpretation of fitness, which is popular in the philosophy of biology. I argue that this interpretation generates all the old troubles with tautologies and circular reasoning. First, let me summarize where we have got so far.

Our main subject has been the status of the phrase "The fittest survive." This phrase has been taken by some as the core of evolutionary thinking. Critics have argued that this will not do, since the phrase is a tautology. To come to grips with the issue, we must know the meanings of "fitness" and "survival." I distinguished two meanings of "fitness": individual reproductive success (pg-fitness) and engineering fitness (e-fitness) representing features of organisms responsible for reproductive success. Considering e-fitness, I argued that reproductive success must not be mentioned in its definition. E-fitness simply represents the features involved, different ones in different cases. Survival can be taken to stand, first, for individual reproductive success (pg-fitness again). Second, it can be taken to refer to the persistence of types in populations.

This provides us with four possible interpretations of "The fittest survive" as a principle of natural selection. First, we can interpret both fitness and survival as pg-fitness. Thus we get a useless tautology. Nowhere in evolutionary biology have I come across this reading of natural selection. It is odd that philosophers have spilt much ink over it. Second, we can interpret fitness as e-fitness, and survival as persistence. This interpretation may be sensible, but we should realize that pg-fitness mediates between e-fitness and survival as persistence. Hence, it is more natural to resort to the remaining interpretations. Third, fitness may refer to pg-fitness and survival to persistence. This yields a sensible principle, provided that it is expanded with *ceteris paribus* conditions. On this reading, the principle of natural selection says that heritable fitness differences cause population change provided that other factors do not interfere. In actual populations, other factors always play a role. Thus, we need a body of more specific principles to explain changes in populations. Our principle of natural selection, on the basis of this interpretation, is best construed as the qualitative claim that fitness differences are an important cause, but not the sole cause, of evolutionary change. Fourth, fitness may stand for e-fitness, and survival for pg-fitness. From the point of view of this interpretation, we are dealing with the existential claim that there are features in which organisms differ such that we get differences in pg-fitness. This claim, as such, has no explanatory force. To get at explanations, we should specify the features involved. Different cases call for different specifications. The claim must therefore be replaced by natural history. We need both the third option and the fourth option. They represent explanation in two stages. Pg-fitness differences together with other factors explain evolutionary population change (pg-explanation). The pg-differences themselves are explained by features of organisms and their environment (e-explanation). Pg-explanations are more general than e-explanations, but it would be misleading to say that they represent applications of a single, general principle of natural selection. A principle with that kind of explanatory power does not exist.

2.6 THE PROPENSITY INTERPRETATION: MILLS AND BEATTY

In the philosophy of biology, discussions about fitness have centred on the so-called propensity concept of fitness—for short, propensity fitness. In line with the analyses in sections 2.4 and 2.5, I argue that this concept is inappropriate because it denotes *relations* between survival and its causes.

The propensity interpretation was introduced in a seminal article by Mills and Beatty (1979). Their definition of "fitness" refers to expected survival. "The $fitness_1$ of an organism x in environment E equals $n =_{df} n$ is the expected number of descendants which x will leave in E" (Mills and Beatty 1979: 275; the subscript df stands for "definition"). This is a definition for individual organisms. Mills and Beatty propose the definition of the fitness of types, $fitness_2$, as the average $fitness_1$ of members of the type (on the basis of this, they also define relative $fitness_2$ in a straightforward way).

The phrase "expected number" in the definition does not represent probability in the relative-frequency sense. Mills and Beatty instead envisage a probability distribution of possible outcomes, in which "the weighting probability for each outcome O_i is just the organism's propensity to contribute i offspring" (Mills and Beatty 1979: 274–275). Several passages in the article indicate that the phrase "propensity to contribute i offspring" in this explication does not stand for pg-fitness. Fitness *consists in* the presence of traits that condition the production of offspring in a given environment (Mills and Beatty 1979: 271); fitness can be *identified with* phenotypic properties causally connected with offspring contribution (Mills and Beatty 1979: 281–282). "Having the propensity to produce i offspring" therefore stands for a *relation* between phenotypic properties (and properties of the environment) and numbers of descendants. On the basis of this interpretation, fitness is a supervenient concept (Mills and Beatty do not use the term) that refers to causes *and* consequences.

Mills and Beatty apparently do not mention phenotypic properties in their definition, because fitness is supervenient on such properties. Now, supervenience is surely incompatible with direct reference to phenotypic properties, but we could also mention them in a more abstract way, as follows: "The propensity fitness of an organism x in environment E equals $n =_{df}$ there are phenotypic properties of x that cause n to be the expected number (in the relative frequency sense!) of descendants that x will have in E." This reading of the definition allows for more clarity because "expected number," in Mills and Beatty's sense, is now unpacked.

Two features of the definition are crucial. First, it has a covert existential component (e.g., "there are properties . . ."). Second, it has pg-fitness in the definiens. *There is a conceptual link between propensity fitness and pg-fitness.*

As argued in section 2.4, this leads to tautologies and circular reasoning, problems that Mills and Beatty are at pains to avoid. Instead of propensity fitness, we should use the concept of e-fitness, which is not conceptually linked to

Problems with Fitness 17

survival. Such a concept would not allow of general theories and explanations. The next section charts connections between propensity fitness and the theme of generality, as elaborated by Sober, in more detail.

2.7 THE PROPENSITY INTERPRETATION: SOBER

Sober (1984, 1993) gives an account of fitness which closely resembles that of Mills and Beatty. I briefly consider some of his arguments because he explicitly considers the subject of generality in defending the propensity interpretation of fitness.

Sober (1984: 47–48) notes that dispositional properties may be described in terms of an associated behaviour and in terms of a physical basis—that is, by reference to effects and by reference to causes. He indicates that fitness represents a special kind of disposition since the physical basis differs from case to case. Fitness is supervenient on the physical basis. A description referring to effects (numbers of descendants), on the other hand, is less problematic. Fitness, then, is taken to be a dispositional property (a propensity) in the sense that "the having of a particular fitness" *means* that there are phenotypic properties (of an unspecified kind) that are associated with a particular expected number of descendants. Thus, Sober appears to endorse Mills and Beatty's view.

According to Sober (1984: chap. 3) fitness does not have a causal role to play with respect to survival. However, he does argue that fitness has an *explanatory* role. Consider the following quotation:

The fact that fitness is a supervenient property also has consequences for the kinds of explanatory roles it can play. Within a single population, fitness is grounded in a set of physical attributes. The fitness differences noted in the model of the sickle-cell trait were based on the properties of resistance to malaria and anaemia. In that example, the concept of fitness explains the same fact that the physical basis explains—namely, the persistence of polymorphism in populations living in malarial regions. [Genes coding for the disease of sickle-cell anemia are not eliminated from populations in regions where malaria is endemic, since persons with the disease have a higher than normal resistance against malaria.]

In this case, it is hard to see how fitness explains something that the physical facts could not explain. . . .

. . . at the same time that the supervenience of fitness predicts that fitness will be explanatory redundant in some contexts, it also predicts that the concept will have a unique explanatory utility in others. Recall that supervenient concepts allow us to generalize over physically distinct systems. They are an antidote to the overly fine-grained analysis that a purely physicalistic vocabulary would impose. The physical basis of heterosis in the sickle-cell system explains the balanced polymorphism we observe there, but in other populations, heterozygote superiority will have a quite different physical basis. Looking at them in isolation, we see that the concept of fitness will help explain the maintenance of the two alleles, but so too will the physical basis of the fitness differences. The vocabulary of fitness and its physical bases will be on an explanatory par, if we take the examples one at a time. But if we want to explain what these various polymorphic populations have in common, the idiom of fitness will be indispensable. [Sober 1984: 83]

Crucial is what the phrase "generalize over physically distinct systems" in this passage means. Sober suggests that propensity fitness is an abstract category that unites a variety of physically distinct properties. However, fitness is not like physical-properties-at-an-abstract-level. It is a *relational* property that connects physical properties with expected offspring numbers.

I would analyze the explanation in the sickle-cell example in a different way, as follows. The relevant physical properties (such as resistance to malaria) associated with particular genotypes *explain* differences in pg-fitness (offspring numbers) between the genotypes (e-explanation). The pg-fitness differences, and the system of inheritance, explain the persistence of particular genotypes in the population, on the assumption that there are no disturbing factors such as differential migration (pg-explanation). Sober, like Mills and Beatty, does not notice that the explanation proceeds in stages.

Now, what happens if we want "to explain what these various populations have in common?" The stage of pg-explanation may be put in a general form without recourse to propensity fitness; pg-fitness will suffice here. Notice that the concept of pg-fitness refers to features (e.g., offspring numbers) that all organisms have. The concept is general, and this allows the formulation of general statements that we can use in explanations. The stage of pg-explanation reveals "what these various polymorphic populations have in common."

E-explanations are less general; their premises refer to specific features such as resistance to malaria. Of course, we can add that *there are* phenotypic properties, different ones in different cases, which are responsible for the observed pg-fitness differences, and this can be expressed by the statement that there are propensity fitness differences or, more appropriately, e-fitness differences. But this is not very informative. Sober's account does not reveal what such explanations would look like. I would propose that we be content with relatively specific e-explanations that share a common pattern. Pg-fitness, not propensity fitness, is a source of generality in evolutionary explanation.

2.8 SOME RECENT DEVELOPMENTS

Disputes over the propensity interpretation have failed to yield consensus. Most authors appear to agree that the general thesis that fitness in the propensity sense causes reproductive success has no empirical content. Yet some argue that the thesis has explanatory value. I shall not analyze all the relevant sources, since the disputes are confusing (e.g., see Nordmann 1990; Ollason 1991; Resnik 1988; Shimony 1989a, 1989b; Sober 1989; Waters 1986; Weber 1996). Instead I focus on Brandon (1990), who to my knowledge has provided the most thorough defense of a nonempirical principle of natural selection alleged to have explana-

Problems with Fitness

tory value. I criticize Brandon's view and subsequently consider criticism of the propensity interpretation by Byerly and Michod (1991).

Brandon (1990: 11) offers the following formulation of the principle of natural selection (PNS): "If a is better adapted than b in environment E, then (probably) a will have greater reproductive success than b in E." Relative adaptation is defined here as relative ability to survive and reproduce. Its meaning is similar to that of fitness in the propensity sense.

Brandon's PNS does confront us with the old tautology problem. The PNS is designed to explain survival by fitness. "Survival" stands here for the reproductive success of individuals. "Fitness" represents the ability to have reproductive success. Hence, the PNS is a tautology. It has no empirical content. As I have argued, a proper reconstruction of evolutionary biology calls for different explications. Fitness in the sense of reproductive success has explanatory force relative to survival in the sense of persistence of types in populations. The tautology problem dissolves under this interpretation.

Brandon grants that the PNS has no empirical content. But its instantiations have empirical content and can be tested. A passage from p. 141 shows how he considers two examples of instantiations:

1. If moth a has darker colored wings than b in (this specific) E, then (probably) a will have greater reproductive success than b in E.

2. If plant a is more tolerant of heavy metals than b in (this specific) E, then (probably) a will have greater reproductive success than b in E.

What does heavy-metal tolerance have to do with dark-colored wings? Without the PNS, not much, Brandon says. The PNS obviously serves to unify diverse examples of natural selection. It provides systematic unification, the gist of scientific explanation.

I agree with Brandon that we should aim at unification to the extent that it is possible. But I would argue that the source of unification must be searched for elsewhere in evolutionary biology. Brandon's two examples have in common that some difference in a phenotype feature causes a difference in pg-fitness (reproductive success). That in itself does not amount to natural selection. We get natural selection in either case, since differences in pg-fitness have the further consequence that, in the absence of interfering factors such as mutation and migration, the difference in phenotype will lead to population change if it is heritable. This is a general empirical claim of population genetics. If you wish, you may call it the principle of natural selection. The principle of natural selection thus conceived is admittedly vague and unspecific. To chart population change in quantitative terms, population genetics needs specific models that account for interfering factors (if they do play a role) for particular types of reproductive system, for interactions between phenotype features, and for much more. Thus, the "principle of natural selection" is replaced by specific models indicating how differences in pg-fitness work out in various situations. The common core of these models is the general idea that pg-fitness differences are an important cause of evolutionary change.

As I have argued, the explanation of evolutionary changes in particular populations proceeds in two stages. First, features of organisms together with environmental factors explain how differences in pg-fitness come about (e-explanation). Second, pg-fitness together with other factors explains population change (pg-explanation). The stage of pg-explanation is more general than the stage of e-explanation. Hence, explanatory unification comes from pg-explanation rather than e-explanation. Brandon's PNS is not a proper source of explanatory unification since its role is limited to e-explanation. The idea that a tautology should be the source of unification in evolutionary explanation is anyhow suspect.

Byerly and Michod (1991) try to capture the dynamics of natural selection by a single overarching idea. They propose that selection is well described by a core equation, which admittedly describes idealized situations:

$(dX_i/dt)(1/X_i) = f(A_i, E)$

Here X_i stands for the number of genotype i in a population; so the left-hand side of the equation represents the per capita rate of increase of this genotype. This rate is expressed as a function both of various adaptive capacities A_i of i and of environmental factors E.

Elsewhere (van der Steen and Voorzanger 1984), I argued in response to a paper on which the approach of Byerly and Michod is based, that a single equation cannot cover evolutionary processes, which are a heterogeneous lot. Byerly and Michod's equation is problematic for that reason. The symbols on the right-hand side, A_i and E, do not stand for variables like "intensity of metabolism" or "temperature," but for *dummy variables* (van der Steen 1991). Even if the equation were applicable to all cases of selection, this would not imply that rates of increase of genotypes always depend on the same variables. On the contrary, in different situations the dummy variables would have to be replaced by different variables.

The equation therefore represents a *metamodel* that covers different context-dependent models. A recent book by Michod (1999) confirms this. He provides an excellent survey of such models. The essence of the metamodel is captured by the following wording. For any genotype in any population of any species, *there are* factors A_i and E, such that the rate of increase of the genotype is a function of these factors. The metamodel as such has no explanatory force. Explanations will be forthcoming only after we specify which factors are operative in the case at hand.

Byerly and Michod's equation is connected with fitness in the following way. They distinguish three fitness concepts. The first concept, A-fitness or general adaptedness, is not a useful notion according to them, since there is no general

Problems with Fitness 21

property of adaptation which enters into causal statements of evolutionary change. Second, Byerly and Michod distinguish r-fitness, a genotype's actual per capita rate of increase. Third, they define F-fitness as the contribution to r-fitness of the genotype's adaptive capacities, reproductive system, and genetic system in interaction with the environment. The equation is meant to cover F-fitness.

Byerly and Michod criticize the propensity interpretation of fitness for the following reason:

The principle of natural selection which Brandon and others formulate in connection with the propensity interpretation is a claim about the effects on a population of differences in fitness (properly F-fitness) over time. However, it makes no mention of the kind of functional dependence of F-fitness on adaptive capacities, environmental variables, and the genetic and reproductive system. An account of natural selection with nontrivial empirical content requires specification of causal connections underlying functional relations among variables which determine F-fitness values. [Byerly and Michod 1991: 13]

This criticism is misleading because Byerly and Michod themselves also do not specify causal connections in discussing F-fitness. The "specification" that their equation provides actually amounts to the thesis that *there are* causal connections that we will have to specify for the context of interest.

Therefore, Byerly and Michod's concept of F-fitness does not differ much from the propensity concept that they criticize. They grant that their concept is supervenient, and that it covers different relationships between features of organisms and (reproductive) survival. These are precisely the features that characterize propensity fitness. Byerly and Michod's analysis obscures that reproductive survival (pg-fitness) has explanatory value. Also, the nature of explanations of reproductive survival itself is unclear on the basis of their reconstruction because reproductive survival is an element in the definition of F-fitness, and because the alleged generality of such explanations is suspect.

The left-hand side of the equation appears to represent changes in pg-fitness. In any particular case, these changes are a function of the values of variables that should take the place of the dummy variables on the right-hand side of the equation. These values will represent features of the organism studied (and features of the environment). We could stipulate that whatever features are relevant in a particular situation will represent fitness (e-fitness in my terminology), but this does not add much to equations covering particular cases. We should grant, though, that specific equations instantiating Byerly and Michod's basic equation have the same form. The equation does not represent general empirical theory, but we could appropriately regard it as an important element of metatheory.

It is interesting that Mills (now named Finsen) and Beatty, who introduced the propensity concept of fitness in their seminal article, are now among the critics of the concept (Beatty and Finsen 1989; for critical comments see Richardson and

Burian 1992). They have two points of criticism. In either case, the problem is that a single measure for expected offspring contribution (reproductive survival) will not suffice to account for evolutionary processes. First, we should distinguish short-term and long-term reproductive success. Short-term success by no means guarantees long-term success. Second, propensity fitness is concerned with probability distributions for offspring numbers, and such distributions are not adequately covered by a single statistical measure such as an expected value (see also Sober in press).

I agree with this criticism, but I would argue that the problem first and foremost concerns pg-fitness. Derivatively, propensity fitness suffers from the same defect because pg-fitness is mentioned in its definition (wrongly so, as I have argued). My own analysis suggests that pg-fitness is the only promising candidate for a general fitness concept, because reproductive survival is a feature shared by all organisms. Other allegedly general fitness concepts (e-fitness, propensity fitness, F-fitness) cannot refer to shared features; this is what makes them supervenient.

Supervenient fitness concepts represent abstract "metafeatures"; they are useful only for "summaries" at the level of metatheory. The domain they cover is best regarded as natural history. The summaries reveal patterns; natural history in biology is not a mere heap of unconnected facts. The use of supervenient concepts in natural history accounts remains misleading, however, to the extent that they further a fruitless search for general theories.

Beatty and Finsen's arguments show that even the generality of any pg-fitness concept may be suspect. Perhaps we need an array of different pg-fitness concepts (De Jong 1994). This would but underline the thesis that general theories are possible only to a limited extent in evolutionary biology.

2.9 CONCLUSIONS

The unfortunate phrase "The fittest survive" has been an important source of confusion in philosophical research aiming to clarify fitness concepts. The phrase is often regarded as the core of evolutionary biology. Some have argued that it is a tautology. Add to this the thesis that the core of scientific theories should not contain tautologies, and evolutionary theory is in trouble.

Considering "The fittest survive," we must notice first and foremost that the term "survival" is ambiguous. It can refer to reproduction of organisms of some type, but also to persistence of types in populations. The tautology problem that has plagued many philosophers of biology, and some biologists as well, dissolves once this distinction is made. Fitness can be defined as survival (reproduction) and still help to explain differential survival (persistence). "The fittest survive" is not a tautology on this construal. It is a general empirical claim that makes sense provided that conditions of applicability are added. Textbooks of population genetics show that biologists regard fitness, in the sense of reproductive success, as one of the causes of survival, in the sense of persistence. They would regard the thesis that differential fitness in this sense explains differential reproductive survival as silly. Philosophers plagued by the tautology problem have failed to understand the nature of evolutionary explanation.

It is convenient to recognize two stages of explanation in evolutionary biology. Factors such as reproduction and migration can explain the survival of types in populations. We can add to this explanation by identifying factors that affect reproduction and migration. I have used the labels of pg-explanation and e-explanation for the two stages. These stages correspond with two kinds of fitness concepts; I have called these pg-fitness, the concept of population genetics, and e-fitness, an overarching concept of engineering fitness. E-fitness can explain pg-fitness, but if we want to have it in that explanatory role, we must take care not to define it as a cause of pg-fitness. If we do that, we generate a tautology problem. The concept of e-fitness covers different features in different organisms. Hence, it cannot be defined by reference to particular features. It is a so-called supervenient concept. E-fitness supervenes on features of organisms. This means that organisms with similar features must have the same e-fitness. But we cannot infer from the fact that some organism has a particular e-fitness what features are involved. The general thesis that e-fitness explains pg-fitness does not have much explanatory force. It boils down to the claim that *there are* features that explain pg-fitness. To get substantive explanations, we must specify which features are involved. Because different features are involved in different cases, e-explanations cannot be general. In industrial melanism, differences in the color of moth morphs help to explain differences in reproductive survival. The explanation of antibiotic resistance in bacteria refers to entirely different features. Both examples represent natural selection. What do they have in common? The answer is not that similar features cause differential survival in the two cases. The commonality is rather that differential survival causes evolutionary change in either case. This amounts to pg-explanation in population genetics. E-explanations enrich population genetics with ecology, but the kind of ecology we need is context-dependent.

Recent discussions have centred on a different supervenient concept of fitness, propensity fitness, which does represent a problematic kind of supervenience due to conceptual links with reproductive survival. Fitness, in the propensity sense, by definition represents features that are responsible for reproductive survival. If we aim to explain reproductive survival by fitness in this sense, we generate the old tautology problem.

Pg-explanations are relatively general; e-explanations, taking the most plausible interpretation, are not, since evolution is brought about by diverse factors. A supervenient concept such as e-fitness which conceptually overarches these factors does allow of abstract metatheories that cover diverse phenomena. However, it will not lead to substantive general theories.

The substance of evolutionary thinking in biology is close to natural history as Darwin presented it. Darwin's original work appears to compare favorably with recent philosophical literature on fitness.

3

Adaptationism

3.1 INTRODUCTION

Evolution could be described, aptly but somewhat one-sidedly, as natural selection generating adaptation. Mutation and other random processes produce variations in populations. The process of natural selection ensures that well-adapted variants survive while other variants are eliminated. Adaptedness amounts to a good fit between organism and environment which ensures reproductive success. Thus, we could expect adaptation to be a widespread feature of the organic world. However, natural selection is not the only force in evolution. Mutation pressure may override weak selection. Genetic drift may have the effect that adaptedness is less than optimal. Pleiotropy—the genetic coupling of features—may cause maladaptive features to share in the success of adaptive features. Hence, adaptation should not be widespread after all.

The thesis that adaptation is widespread has been called adaptationism. Adaptationism in this substantive sense is false. In a different, heuristic sense, adaptationism is a research strategy aiming to uncover adaptation whenever it exists. The strategy is sensible, but the heuristic may deteriorate, since it is easy to invent stories that postulate an adaptation where none exist. Gould and Lewontin (1979), in a famous paper, have criticized tendencies to overvalue adaptation, and selection as a source of adaptation. One of their points is that links need not exist between selection and adaptation. Selection does not always lead to adaptation, and adaptation is not always caused by selection.

Substantive adaptationism, interpreted as the strong thesis that adaptation resulting from selection is widespread, is false, but a weaker variant may make sense. Natural selection might be so potent as a force that other forces deserve to be ignored in model-building. In section 3.2, I consider this variant of adaptationism as portrayed by Sober (1987), and I criticize one of Gould and Lewontin's arguments that adaptation need not result from selection. They argue that physiological adaptation, by which they mean *differences* in adaptation due to the environment, does not result from selection. That is true, but we should notice that adaptation thus becomes a matter of comparison. Features may represent a physiological adaptation under some comparison, and a genetic adaptation due to selection under a different comparison. Since the choice of comparisons is up to us, we can always construe a feature as a physiological or as a genetic adaptation. Thus the thesis that adaptation need not result from selection becomes a trivial conceptual matter. The need to specify comparisons makes sweeping claims about adaptation-in-general impossible. We must instead resort to natural history.

Without specification, the notion of adaptation is a meaningless placeholder. Features are not adaptive, period. They can only be called adaptive relative to alternative features, relative to an environment, against the background of the entire phenotype. In technical terms, adaptation is not a one-place predicate but a many-place predicate. To make sense, it must not be linked with one item (a feature), but with a variety of different items.

One of Gould and Lewontin's arguments that selection need not produce adaptation, about which they are brief, has more force. Suppose that a population is invaded by a mutant that has a higher reproductive success than an existing variant and is similar to it in other respects. The mutant will outcompete the existing variant, but this need not result in any change of population density. Gould and Lewontin regard this as an example of selection without adaptation. I agree, if adaptation is taken to mean the solution of a problem posed by the environment. I would characterize the situation as selection without environmental selective agents. In section 3.3, I elaborate this theme, which deserves scrutiny because many biologists and philosophers of biology assume, implicitly if not explicitly, that selective agents play a role in all processes of natural selection.

3.2 ANALYZING RIDDLES

Let us take the thesis of adaptationism to mean that natural selection is the overwhelmingly most powerful force of evolution. Sober (1987) has suggested that we could test the thesis of adaptationism by modeling selection without taking mutation, drift, and pleiotropy into account. If our models generally generate predictions that fit well with the situations modeled, we are entitled to conclude that adaptationism is confirmed. We know that mutation, drift, and pleiotropy affect evolutionary processes, but our predictions would then indicate that their force is limited compared to the force of selection.

The strategy suggested by Sober seems to be reasonable. We can assess the relative importance of selection, mutation, drift, and pleiotropy by appropriate tests. Considering the role of these factors, Gould and Lewontin's worry appears to be that biologists at times endorse adaptationism on the basis of just-so stories without conducting proper tests. Their anti-adaptationism thus amounts to the useful warning that we must be critical.

Gould and Lewontin also argue that selection need not result in adaptation, and that adaptation need not result from selection. This implies that adaptation is not defined as resulting from selection. On the basis of such a definition, which has some currency in the literature, adaptation without selection would be impossible, for logical reasons. Sober appears to presuppose that adaptation is defined in some such way. Otherwise it would be unclear why he characterizes a thesis about the predominance of selection as adaptationism rather than selectionism.

Gould and Lewontin first point out that mere reproductive success may lead to the proliferation and fixation of a mutation. That would amount to selection without adaptation. I have called this selection without selective agents. I regard this as the most salient point made by Gould and Lewontin. In section 3.3, the issue is analyzed in detail.

Second, Gould and Lewontin consider phenotypic plasticity, which they regard as adaptation without selection. Some organisms show physiological adaptation in that they develop different, appropriate responses in different environments. For example, sponges and corals develop different, good phenotypic designs under different flow regimes. Gould and Lewontin state that these adaptations are not heritable, though the capacity to develop them presumably is. They do not result from selection. We must distinguish physiological adaptations from genetic adaptations that may result from selection.

I feel uneasy about this example, because all features of organisms—physiological adaptations included—result from genetic and environmental factors. Gould and Lewontin's assertion that physiological adaptations are not heritable has to do not with features taken in isolation, but with a comparison of features in different environments. In phenotypic plasticity, differences in features are caused by an environmental difference, not a genetic difference.

Suppose that an organism has feature A in one environment and feature B in a different environment, the difference being nongenetic. Now let us add to this scheme another organism, which develops C in the first environment, the difference between A and C being genetic. This allows us to say that feature A is a genetic adaptation after all, not a physiological adaptation. Thus, the attribution of physiological adaptation or genetic adaptation to individual features of organisms is context-dependent. Adaptation is a matter of comparison, and a given feature can always be subjected to different comparisons that involve different forms of adaptation, or adaptation versus nonadaptation. A feature can only be an adaptation relative to a particular environment, compared to particular other features, against the background of the entire phenotype. Thus, "adaptation" is a many-place predicate because it is meaningful only if it is connected with several items, not merely with features in isolation. It is a placeholder concept with different shades of meaning in different contexts. This makes general claims about adaptationism suspect. The need to specify comparisons calls for a natural history approach.

Third, Gould and Lewontin consider cases that involve selection and adaptation where no selective basis exists for differences among adaptations. For example, species of the West Indian land snail genus *Cerion* that live in exposed places have thick shells. This is presumably an adaptive feature resulting from natural selection. Different species have different shell designs, and these differences may not result from selection. They may represent historical contingency rather than optimal design.

Comparisons again! I regard the comparison made here as irrelevant. Suppose the genus *Cerion* would consist of a single species. Then we would also assume that some aspects of shell form are due to selection, while other aspects are a matter of historical contingency.

Concerning adaptationism, historical contingency is immaterial, since all features of organisms in part reflect historical contingency. Natural selection, after all, has to work on variations produced by random mutations. No adaptationist would deny that. It is futile to ask whether the mutations are more important or less important than the selection process working on them.

Fourth, selection may result in adaptive features, in a process that generates as a byproduct other features that may allow of profitable utilization. Gould and Lewontin make the point here that we run the risk of seeing design in nature where none exists. This is a useful reminder that we must be critical.

To the list of Gould and Lewontin, we should add that natural selection, most notably frequency-dependent selection—leads to maladaptive traits in some situations (Hartl and Clark 1989; Michod 1999).

Gould and Lewontin's article rightly warns us against excesses in adaptationism resulting from just-so stories without tests. But some of their arguments are odd. Sober's suggestions show that critical forms of adaptationism make sense, but he appears to overlook some problems with adaptationism since he departs from a narrow definition of "adaptation." As indicated, I regard selection without adaptation—selection without selective agents—as the most important point made by Gould and Lewontin. They only discuss it briefly, and it has received little attention afterwards in the literature. So let us consider selection without selective agents in more detail.

3.3   SELECTION WITHOUT SELECTIVE AGENTS

Differential reproductive survival of entities (of phenotypes and genotypes, and thereby genes) results in evolutionary change. The process resulting in the change is called natural selection. This easily evokes the idea that natural selection is a cause of evolution. So far, I took that idea for granted. But its meaning is by no means clear. In what sense is natural selection a cause?

From one interpretation, "natural selection" simply *means* "differential survival." Thus, it may help to explain evolution, but only partially so since the causes of differential survival are not specified. On a different interpretation, natural selection *causes* differential survival, and thereby evolutionary change in populations.

The latter interpretation also calls for further specification. Natural selection is not a factor like temperature or a process like temperature influencing differential survival. If a population has two phenotypes, one of which cannot survive at high temperatures, and if the population comes to be exposed to such temperatures, then high temperature will cause one phenotype to go extinct while the other phenotype survives. The factor of temperature then causes the extinction of one phenotype, while allowing the survival of a different phenotype.

True, we can describe this situation by stating that selection caused evolution through differential survival of types. But that is an elliptical manner of speaking. Selection is not a cause over and above temperature and other environmental factors, or a process over and above temperature and other environmental factors effecting differential survival. So, any thesis to the effect that selection caused evolution through differential reproductive success in a particular case should be unpacked as the thesis that *there are* factors such as temperature that, together with phenotype features, are causally responsible for the evolutionary change. Selection, in brief, is a placeholder concept.

Temperature, in the example, would be an environmental factor that causally influences the fitnesses of different phenotypes in different ways, such that evolutionary change ensues. In line with common parlance, I refer to factors working in this way as *selective agents*.

The equations of population genetics—according to many researchers, the heart of evolutionary theory—do not inform us about particular phenotype features and selective agents. They deal with selection in a more abstract way. To get at substantive causal pictures, we have to supplement the equations with ecology.

The notion of natural selection, like the notions of engineering fitness (see chapter 2) and adaptation (see section 3.2), is best regarded as a kind of placeholder for factors and features that do the causal work in evolutionary processes. These factors need to be specified to get at full-fledged explanations. Causal explanations in evolutionary biology thus thrive on natural history, not general laws (see also chapter 2).

We are always dealing here with compound processes. At issue is a comparison of types with different fates. This invites the question of whether we can decompose selection processes and regard them as aggregates of processes involving separate types. In the temperature example, decomposition is possible in that the extinction of one type and the survival of another type are distinct processes. True, if the population exhibits sexual reproduction, interactions among the processes exist, but this affects ultimate outcomes only in special cases such as heterozygote superiority. Since sexuality may detract from some points I want to make, I assume for the moment that we are dealing with asexual populations.

A comparison of types acquires a more fundamental meaning if processes are not simply aggregative—that is, if the differential survival of types is due in part to interactions between types. Suppose that food rather than temperature is a salient factor. If two phenotypes differ in foraging efficiency, and if food is in short supply, then it is conceivable that efficient foraging in one type is a causal factor contributing to the demise of another type. We are dealing then with differential survival due to intraspecific competition. Darwin's catchphrase "struggle for existence" does evoke the picture of competition.

Purely aggregative selection should indeed be rare. For example, in the temperature case, the extinction of one phenotype will allow the other phenotype to reach higher densities if total population density is regulated in any way. Thus the effect of temperature on one phenotype causes another phenotype to increase in density. This is a kind of causal interaction. The interaction in this case is more indirect, though, than in the foraging example.

Equations of population genetics do not tell us which environmental factors and phenotype features are responsible for differential survival. Therefore, they do not suffice to explain the outcome of particular selection processes. Full-fledged explanations should specify operative factors and features. Thus, general, abstract models of population genetics will have to be supplemented with "natural history" in the domain of ecology that conveys what factors and features are operative in the case at hand.

We should beware thinking here that some natural history description should provide the best causal account of any particular selection process in the field or in the laboratory. Causal attributions are context-dependent because they presuppose comparisons. What caused a particular phenotype A to go extinct, while another phenotype B survived in a particular environment E? On specifying the context in this way, we can point to particular phenotype features as salient causes. What caused this phenotype A to go extinct in environment E, whereas it would have survived in another environment? Now the salient cause becomes some environmental factor (e.g., temperature) or an array of environmental factors.

In elementary models of population genetics, the envisaged comparison is generally between types in a given environment. Thus, they need not entail that environmental factors are causally implicated in *differential* survival. That is, the models do not allow us to assume that the environment affects different types in different ways.

It is commonly assumed that environmental factors do contribute to fitness differences as modeled in elementary population genetics (differences in pg-fitness: see chapter 2). This suggests that *selective agents* (environmental factors that make a difference causally regarding fitnesses of different types) are always present, although we do not need to know about them. But models that depict the fate of types with different fitnesses in a given environment do not presuppose that selective agents thus conceived exist.

Imagine two phenotypes that are characterized by a physiological difference in the efficiency of food digestion. Such phenotypes may well show similar differences in fitness in any environment they may occupy. Although the environment is necessarily a causal background of fitness in either type, it need not make a causal difference in any comparison of phenotypes (genotypes, genetypes) we may wish to make. Gould and Lewontin (1979) made this point in their defence of the thesis that selection need not produce adaptation.

The example of food digestion is not realistic, in that it disregards population regulation. Let us become more realistic. Consider again two different phenotypes that differ in the efficiency of food digestion such that their relative fitnesses are at different, constant levels in any environment they may inhabit. If some environmental factor that does not discriminate between the phenotypes (say, predation) ensures that total population density has a fixed value, the phenotype with the lowest fitness will go extinct due to this. Absolute fitnesses are density-dependent (and also frequency-dependent) in this case; relative fitnesses are not. Thus predation is a causal factor which contributes to differential survival, but it affects individuals with different phenotypes in the same way. That is, the probability that a particular organism dies as a result of predation does not depend on its phenotype.

This example is interesting since predation is here a random process in that individuals eaten by predators at any particular point in time constitute a random sample from the prey population. Yet predation is an essential part of the selection process since without it no phenotype would go extinct. Causally, it makes a difference in that the outcome of selection would be different without it. But it does not affect different phenotypes in different ways. Thus, it would be odd to call it a *selective agent*. I shall use the term *contributive agent* for factors that influence selection in this way. Temperature in the earlier example is a selective agent; predation in the present example is a contributive agent (Analogously, we may distinguish between *selective phenotype features* and *contributive phenotype features*. I disregard this possibility.)

If no selective agents are operative, contributive agents are causally implicated in the selective elimination of phenotypes (and thereby genotypes and genes) even though they do not differentially affect fitness.

Different categorizations of selection processes have been distinguished in evolutionary biology (e.g., stabilizing selection, disruptive selection, frequency dependent selection, and so forth). My analysis indicates that we should add a new categorization to extant distinctions, since the environment may play different roles in selection processes. We should distinguish between selective agents and contributive agents (and combinations thereof), and we should dispense with the overt or covert assumption that selective agents are operative in all selection processes.

It is amazing that biologists habitually tend to assume that selective agents must exist whenever selection occurs. Thus, Endler (1986) in his well-known book has a table (pp. 129–153) with an extensive survey of studies uncovering natural selection, which has a column headed "selective agent." For each study, Endler either has particular environmental factors as entries in this column, or lists selective agents as "unknown" (but apparently presumed to exist).

Philosophers of biology have also assumed that selective agents are necessary for selection to occur. Thus it seems that Brandon (1990), in his equally well-known book, implicitly endorses this view. The following passage shows how he introduces the notion of natural selection. It indicates how easy it is for seemingly innocuous, covert assumptions to become entrenched in evolutionary thinking.

Suppose that in a given population of organisms there is directional selection for increased height. To say that there is directional selection for increased height is to say that taller organisms have (or tend to have) greater reproductive success than shorter ones, that is, that reproductive success is an increasing function of height. The ecological reasons for this can be indefinitely varied. For instance, in one ecological setting taller plants may receive more sunlight and so have more energy available for seed production. In another setting, taller animals may be more resistant to predation. Also, differences in reproductive success can result from differences in fecundity (as in the first case) or from differences in survivorship (as in the second case) or from still other causes (e.g., differences in mating ability). [Brandon 1990: 4]

Brandon does not explicitly say here that selective agents are necessary for selection to occur. But the assertion that "the ecological reasons for this can be indefinitely varied" appears to presuppose that such reasons always exist. I would note here that, even though the environment necessarily contributes to selection, *differences* in reproductive success need not have any ecological reasons.

A crucial distinction in Brandon's reconstruction of evolutionary theory is between the external environment (the sum total of external factors influencing survival and reproduction), the ecological environment (environmental features affecting the organisms' contributions to population growth), and the selective environment (aspects of environment measured in terms of the relative actualized fitnesses of different genotypes across time and space) (pp. 47–49). Revealing is the following remark about the selective environment: "One step [in the process of evolution by natural selection] is the selective discrimination of different phenotypes. This process occurs within an environment, that is, the relative adaptedness of the phenotypes depends on the particular environment" (p. 50). ("Adaptedness" is Brandon's term for "fitness.") Here the expression "depends on" again suggests that selective agents are involved. The very phrase "selective environment" also suggests this.

In an account of explanation in evolutionary biology, Brandon presents a similar picture of the role of the environment. He argues that an "ideally complete adaptation explanation" should have five components: (i) evidence that selection has occurred, (ii) an ecological explanation of the fact that some types are better adapted than others, (iii) evidence of heritability, (iv) information about

Adaptationism

population structure, and (v) phylogenetic information. I would argue that item (ii) may be irrelevant in some cases, because selective agents need not be involved.

Closest to my own view is the analysis of selection by Lennox and Wilson (1994). They argue that, in the present century, many biologists and philosophers of biology have decoupled the struggle of existence—a key element in Darwin's theory—from natural selection. Selection has come to mean mere differential survival of types leading to a change in population composition. But in Darwin's view, selection involves a struggle for existence—that is, checks on population growth—in addition to this. In the view of Lennox and Wilson, it is misleading to oppose selection in the new sense to genetic drift. Instead, we should distinguish three types of evolutionary process (p. 76):

1. *Darwinian selection:* Changes in genotypic (or phenotypic) frequencies due to differential survival and reproduction of phenotypic variants better able to deal with environmental constraints.
2. *Changes due to differences in reproductive fitness:* Changes in genotypic (or phenotypic) frequencies due only to differences in intrinsic rates of reproduction among reproducing genotypes (or phenotypes).
3. *Genetic drift:* Changes in genotypic (or phenotypic) frequencies due to chance fixation and spread of variant genotypes (or phenotypes).

In the second type of process, "the environment will be causally neutral with respect to the relative reproductive success [of the different genotypes or phenotypes]. In such cases, the environment has an 'indirect effect' on the reproductive success of the members of the population, since it serves only as a stable causal background against which the different reproductive capacities of the organisms are realised" (p. 77). Lennox and Wilson assume that checks on population growth (Darwin's struggle for existence) involved in selection entail differential effects of the environment on different genotypes or phenotypes. I have shown by examples that this assumption is not necessarily valid. Hence the distinction between Types 1 and 2 should be replaced by the distinction between selection involving selective environmental agents and selection involving contributive environmental agents. The latter category comprises Type 2 and part of Type 1. Checks on population growth exist in all these processes in the sense that the environment, at the very least as a causal background, influences growth rates. Checks in a stronger sense have the further effect that the population density stays below a particular ceiling. The distinguishing mark of Type 2, which can only exist over limited time-spans, is that strong checks are absent in this case. In this respect, it is admittedly a special case.

Models of population genetics chart consequences of fitness differences (differences in pg-fitness, in the terminology of chapter 2). They do not tell us what features of organisms and what environmental factors are causally responsible for these differences. My analyses show that the environment, together with phenotype features, can play different roles in selection. So a given model of population genetics may be compatible with different causal scenarios. The models are anyhow silent about essential causes. Therefore it is natural to wonder if they can have any explanatory role. I argue that they do have such a role, albeit a limited one. I contrast my view with the views of Sober.

Sober (1984: 50–51, 59) argues that the laws of population genetics are consequence laws that describe results of fitness differences. In addition to them, we need source laws that describe the causes of fitness differences. Source laws are in the province of ecology. They are hard to come by since the causes of fitness differences are different in different cases. Sober also argues that fitness, though potentially explanatory, cannot be a cause. The overall fitness of an organism is determined by many different features, and it should not be regarded as a cause over and above such features. Consider heterozygote superiority (pp. 83–84): we can causally explain how the superiority works out in particular cases by appealing to the physical basis of fitness differences. Sickle-cell anemia is an example. Persons who are heterozygous for the sickle-cell trait have a high fitness because the condition affords some protection against malaria. Thus we can give a perfectly straightforward explanation for the phenomenon that sickle-cell anemia, though disadvantageous, stays in populations subject to malaria. In this case, the causally relevant features are resistance against malaria and suchlike. The high fitness of heterozygotes is not a separate feature that adds to this in a substantive way. Fitness is not an additional operative cause at a higher level. However, talk of fitness is explanatory here because it unites in a general way what diverse cases of heterozygote superiority have in common.

My analysis of explanation is different (see also chapter 2). I agree that fitness (as understood in population genetics) cannot sensibly be regarded as a cause of reproductive success. But I would add that fitness definitely is one of the causes of population change. Let us simplify and assume that "fitness" is defined as "(expected) reproductive success." Now, changes in populations are determined by the reproductive success of different types, *together with factors such as migration and mutation*. Models of population genetics show how these factors combine to produce change. Fitness is among the salient causes. The fact that fitness in turn is determined by all sorts of features does not imply that it cannot be a cause. Sober is here confusing fitness as a cause with causes of fitness.

Considering the example of heterozygote superiority, we can legitimately state that fitnesses are among the causes that explain the persistence of sickle-cell anemia in certain areas. However, we would also like to know how fitnesses themselves are to be explained. That is where Sober's physical basis of fitness comes in. We now have to resort to causally relevant phenotype features and environmental factors which are different in different cases of selection.

A full-fledged explanation of (short-term) evolutionary processes proceeds in stages. One stage is covered by population genetics. Here we can give relatively general explanations showing how fitness differences, mutation, migration, and other factors causally influence population change. Next, we can explain how fitness differences come about. That is more a matter of ecology—under the present demarcation of disciplines, which we may well deplore. This stage of explanation cannot be very general. In part, therefore, evolutionary explanation cannot do without "natural history" (see also the comments on engineering fitness—e-fitness—in chapter 2).

Sober further argues that the concept of selection, unlike fitness, can feature in causal attributions. To show this, he introduces the illuminating distinction of two notions, selection *of* and selection *for* (pp. 97–102). Due to selection, all sorts of features may come to be established in populations, while other features are eliminated. But this does not imply that selection for these features has occurred. For example, due to pleiotropy selectively neutral or even disadvantageous features may get established simply because they are linked with selectively advantageous features. We then have selection *of* the neutral or disadvantageous features and selection *for* the advantageous features. Selection for phenotype features takes place if the features are causally responsible for reproductive success.

I agree with this analysis, but I would make the additional distinction of selective versus contributive agents, and I would expand the analysis to emphasize again the crucial role of natural history in evolutionary explanations. The term "selection for" represents the idea that *there are* environmental factors that do the selecting (selective agents). Consider the well-known example of industrial melanism in moths. In Sober's terminology, we can state that selection for dark morphs explains differences in fitness between light and dark morphs. But that is not much of an explanation if we do not say why selection for dark morphs exists. We would not know that this kind of selection *for* is happening, unless we knew that it is a matter of protective coloration that shields dark individuals against predation. To get at a satisfactory causal picture, we need to know that selection *for* particular features exists, but in addition to this we need information about selection *by* particular selective agents for these features. This information generates natural history, not general theory. If no selective agents exist, we need to know how contributive agents affect particular features.

Brandon (1990) is more explicitly concerned with natural history. According to him, the core of the theory of evolution by natural selection is the principle of natural selection, which, as we saw in section 2.8, he formulates as: "If a is better adapted than b in environment E, then (probably) a will have greater reproductive success than b in E" (p. 11). Brandon holds that this principle is empirically empty. However, the presuppositions of the principle are empirical (pp. 144–153). One presupposition, for example, is that fitness differences do exist in populations. Brandon further argues that, if the principle is made concrete for particular cases, we get a perfectly legitimate empirical picture (e.g., the adaptation explanations I mentioned). He works this out in an informative case study of heavy-metal tolerance in plants, in which an expansion of population genetics with ecology yields a picture that deserves the label of natural history (not Brandon's term).

Brandon rightly notes that, if fitness is defined as reproductive success, it cannot at the same time explain reproductive success. His principle of natural selection is indeed empirically empty. But population geneticists are not concerned to explain with their models the reproductive success of types. They want to explain changes in the frequencies of types in populations. As I argued, fitness as reproductive success is one of the causes of these changes; there are other causes such as mutation, migration, and the like. This allows of general causal explanations of population change. The explanations can be expanded by ecological information about the causes of fitness differences, as Brandon shows by his case study of heavy-metal tolerance.

According to my reconstruction, population genetics is general and empirical. But the full causal story about selection calls for ecology also, and it should be replete with natural history assigning context-dependent roles to particular selective agents and contributive agents.

3.4 CONCLUSIONS

Natural selection is widely regarded as an important cause of evolution which leads to adaptation. Gould and Lewontin (1979) have argued that we should not regard adaptation as widespread, and also that selection and adaptation need not be tightly linked. They interpret adaptationism as the thesis that adaptation is widespread, and that it results from selection. They rightly argue that we should not uncritically attribute adaptedness to features of organisms and assume without evidence that selection is responsible for all adaptations.

Gould and Lewontin's view that adaptation need not result from selection is problematic. They argue that physiological adaptation, interpreted as adaptive differences in features due to the environment, is not caused by natural selection. Now, physiological adaptation is a matter of comparison. A feature can represent a physiological adaptation under one comparison, and a genetic adaptation due to selection under a different comparison. The choice of comparisons is ours. Hence, the general thesis that physiological adaptation does not result from selection is uninformative. The fact that we need a comparative approach reinforces a point made in chapter 2 that evolutionary biology thrives on natural history rather than general laws of nature.

The conception of selection as one of the causes of evolution is potentially misleading. Selection is not a factor like, say, temperature. It functions as a placeholder for particular factors, different ones in different situations. Selection involves differential reproductive survival of phenotypes and genotypes. Particular environmental factors and features of organisms are causally responsible for the differences in survival. Selection does not have a causal role in addition to this. Models of population genetics do not specify the causal items responsible for differential survival. Such items are in the province of ecology. Full-fledged explanations of particular evolutionary changes therefore need to rely on population genetics enriched with ecology. The explanations typically take the form of natural history.

Considering the role of environmental factors, we must distinguish between selective agents and contributive agents. Selective agents are environmental factors that are causally responsible for differences in reproductive success. Natural selection need not always involve selective agents; this is indeed an important point made by Gould and Lewontin, in different terminology (selection without adaptation), in their criticism of adaptationism. The environment may also contribute to the differential survival of types without being causally co-responsible for fitness differences. I suggest that we use the label of contributive agent for environmental factors with this role. The distinction of selective agents and contributive agents calls for a new categorization of selection processes in addition to existing ones.

4

The Chimera of Optimality

4.1 INTRODUCTION

Natural selection is often portrayed as a process that generates optimal traits in organisms. I argue in this chapter that the thesis of optimality is problematic because general notions such as "optimality" are unclear. Considering examples of selection processes, we should investigate *in what respects* optimization takes place. This inevitably leads to natural history, not general theory, as I argued on other grounds in the previous chapters.

I focus on optimal foraging theory (OFT) as a variant of optimality theory. The origin of OFT lies in the 1960s, when the idea emerged that behavior, like morphology, is optimized in the course of evolution. The earliest papers are those by McArthur and Pianka (1966) and Emlen (1966) on prey selection. The literature on OFT has subsequently grown exponentially (for a review see, for example, Schoener 1987). OFT became successful, but as it expanded its methodological problems increased as well.

The basic idea of OFT is that the foraging behavior of animals is optimized such that fitness is maximized. But this must be qualified. Unrestricted optimality is impossible. Lizards cannot fly, so they will not catch prey up in the air, however profitable these might be. That is why we need the notion of *constraint*. Constraints are restrictions that define the space where the optimal feature must be found. The options available within this space constitute the *strategy set*.

OFT models evaluate the options within constraints through a *currency* variable. A commonly used currency is the net intake of energy per time unit. The optimal option should maximize energy intake, to the extent that fitness (reproductive success) is thereby maximized.

Crucial concepts of OFT are hard to define. The concept of optimality itself is a prime example. Consider the idea that unrestricted optimality is impossible. At first sight this idea captures a sound empirical statement. If so, then the notion of unrestricted optimality, or optimality-in-general, should be meaningful. But, in point of fact, the notion hardly makes sense. Look at the gulls that fly over your head. If I were to ask you if they are performing an optimal behavior, your response would presumably be that I should ask a more specific question. Questions about optimality become meaningful only after the specification of constraints, strategy sets, and currencies. "Optimality," like "fitness," "adaptation," and "selection," is a placeholder concept with different meanings in different contexts. More specifically, it is a many-place predicate, like "adaptation" (see chapter 3).

I focus on constraints. First, in section 4.2 I consider some problems with the generality of OFT, which are more fully covered in Haccou and van der Steen (1992).

Under a full specification of the features that constrain foraging, behaviors should be optimal by definition. Thus we get a meaningless general thesis of optimality. OFT has indeed been regarded as untestable because invalid models can be modified by new constraints so as to fit the evidence (Pierce and Ollason 1987). I reject this charge. I argue that OFT has a general core that is testable—indeed, well confirmed. But the core of OFT does not have much empirical content, and it should not contain troublesome notions such as optimality and constraint. Specific models are also testable, but their explanatory force is limited. They serve descriptive purposes in that they identify constraints that allow of limited forms of optimality in particular animals. Rejection of a model only indicates that the descriptions it provides are not realistic—for example, because it disregards particular constraints.

4.2 GENERALITY, UNIVERSALITY, AND TESTABILITY

The methodological criterion of generality has two different meanings. First, a statement can be general in the sense of universal. A statement is universal if it contains a universal quantifier (e.g., "*All* animals have . . .") and does not mention particular places, times, or individuals. Second, generality stands for the opposite of specificity; I reserve the term "generality" here for this notion. For example, the statement that all mammals are homeotherms is more general (less specific) than the statement that all mice are homeotherms, since it covers more items. Notice that universal statements can be highly specific.

Because the general concept of optimality is not meaningful, there is no place for it in a general core theory of OFT. What, then, should a general core theory look like? I would argue that the basic, qualitative general tenet of OFT is that, given a set of environmental characteristics and features of organisms, their foraging should allow them to have descendants in the long run. Behavior may deviate slightly from this rule, but this is only acceptable if the deviations can be explained by, for example, recent environmental changes. The basic tenet is general and universal.

The core theory is testable if certain types of behavior are incompatible with it. It is easy to show that this is indeed so. For example, consistent energy minimization is at variance with the theory, since animals performing that behavior have no descendants in the long run. If there were to be individuals consistently showing minimization of energy intake, the core of OFT would be invalid. Notice that this behavior can be described with an optimality model. The point is that the currency—minimization of energy intake—violates the general idea of OFT. Examples of unacceptable constraints also exist. For example, the constraint that an animal must leave a patch after having eaten one prey item would be unacceptable when the distance between patches is so large that it would starve before finding the next patch.

The core of OFT implies that some models are unacceptable, whereas others are acceptable. OFT therefore has empirical content, and it is testable. The fact that, in the examples I gave, individuals could not even have descendants in the short run supports the thesis that evidence against the general core of OFT is possible. Many kinds of unacceptable currencies and constraints are simply not considered in the literature. That is understandable because it is obvious how we should think about them. Those who argue that OFT is not testable are overlooking the obvious.

Those who defend OFT against the charge that it is untestable often comment that the thesis of optimality is not under test and that only specific models are tested (Maynard Smith 1978; Stephens and Krebs 1986). This is confusing since it presupposes a meaningful general thesis of optimality. No such thesis exists. Defenders and critics of OFT are caught here in the same conceptual trap.

The core of OFT excludes certain classes of behaviors. Beyond that it is not informative. For substantive information, we must resort to specific models. It would be nice if we could expand the core of OFT with models containing laws that are universally valid for all animals, but the signs are that this is impossible. Universal laws are scarce in biology.

In the philosophy of science, some researchers have argued that problems with universality dissolve under the so-called semantic view, which considers theories as ideal systems without empirical content (see also chapter 2). From this point of view, empirical matters are covered by theoretical hypotheses that state that some empirical system is identical to an ideal one (for a review of applications in biology, see Thompson 1989).

Beatty (1980) argued that the semantic view solves the problem of universality in biology, because it construes laws as part of ideal systems. For example, we could regard Mendel's laws as universal statements that describe a particular ideal system of heredity. Empirical matters concerning heredity would then be described by hypotheses saying that *particular* features in *particular* populations behave in accordance with the ideal system. Laws at the level of ideal theory thus would not have exceptions.

This view has been applied to OFT by Stephens and Krebs. They argue that we should not expect universality in models of optimal foraging:

Optimality models fail the "physics test" because they do not specify the range of their own validity. The models of Chapters 2 and 3 say what rate-maximizing foraging should be like, but they do not say that all animals meeting criteria X, Y, and Z will be rate maximizers, as a physical theory would. Optimality models specify types of systems, and as such they are yardsticks against which to compare nature; they are not claims about what nature must be like ... [Stephens and Krebs 1986: 212–213]

However, as Sloep and van der Steen (1987) have objected, the problem of universality is simply displaced in this way. The old ideal of universality should now apply to theoretical hypotheses rather than to old-style laws of nature.

I would not endorse the view that theories and models in biology should always be universal. Biology contains much useful nonuniversal "natural history" that cannot be replaced by universal claims (van der Steen and Kamminga 1991). Yet we should always try to move in the direction of universality, if that is possible. Concerning OFT, it is unclear as yet how far we will be able to get. Meanwhile, we have valuable knowledge about optimal foraging in the form of natural history.

What form should universal claims concerning OFT models take? I think that it is best to start with schematic statements such as: "If conditions C are satisfied, and if fitness of animal A is maximized by maximization of quantity Q, then A will behave in such a way that Q is indeed maximized." Here, Q represents a currency, and C covers constraints and other interfering factors. C, Q, and A represent variables that can generate different specific models. A specific model will describe, say, the maximization of a *particular* quantity for a *particular* taxon, and it will recognize constraints. As such it will not be universal in intent—though it may be valuable if well confirmed. To arrive at universal claims concerning such models we will have to be specific about C and about the postulated relation between Q and fitness.

The danger is that universal claims are made untestable *in practice*. This may occur through, for example, the following moves. If a model is contradicted by data, it is simply *assumed* that there is an unidentified constraint. If a model fits the data, it is simply *assumed* that maximization of the currency chosen amounts to the maximization of fitness. I discuss constraints extensively in the rest of this chapter.

4.3 CONSTRAINTS AND FREE WILL

In literature on optimal foraging and, more generally, literature on evolution, constraints are commonly understood as constraints on adaptation. "Adaptation" stands for a good fit between organism and environment which enhances fitness in the sense of expected reproductive success.

The Chimera of Optimality

Explicit definitions of "constraint" are seldom given. The following definition presumably captures the commonly intended meaning of "constraint" in OFT. Feature C is a constraint for organism O if and only if O has C, and there is at least one feature F such that C prevents O from having F, while F would be an adaptive feature for O (relative to O's environment). Constraints on optimality can be construed as a special case of constraints on adaptation. Such constraints prevent O from having a feature F that would be maximally adaptive for O (i.e., a feature F that would maximise fitness).

We can generalize the notion of constraint on adaptation (*A-constraint*, for short) by deleting the restriction in the last part of the definition. Thus, generalized constraints (*G-constraints*, for short) are defined as follows. Feature C is a constraint for organism O on feature F if and only if O has C, and C prevents O from having F. Amundson (1994) has introduced a slightly different distinction. He rightly pointed out that constraints *on form* in developmental biology are more inclusive than constraints on adaptation. His constraints on form are a subset of my G-constraints.

The notion of A-constraint used in OFT is not as straightforward as casual inspection may suggest. For one thing, it refers to the concept of adaptation or, alternatively, fitness, and these concepts are by no means unproblematic. I disregard problems with these concepts since they have been exposed in previous chapters. I focus here on different problems with the concept of constraint.

Fundamental problems attach to the seemingly innocuous notion of (im)possibility that figures underneath the constraint concept. Constraints on optimal foraging prevent the organism from performing some behaviors, while allowing other behaviors. Constraints are thus supposed to provide a distinction between possible behaviors (which constitute the strategy set) and impossible behaviors. The aim is to investigate whether animals select the best strategy from the set of possible behaviors. What do we mean when we say that an animal refrains from certain possible behaviors? A natural way to make sense of this is to rely on the notion of choice. The animal could perform the behaviors, but it chooses not to do so.

It should be possible in principle to explain the choices that animals make. The choices will be causally determined (perhaps strictly, at least probabilistically) by dispositions and/or internal states of the animal (together with environmental conditions and cues). If an animal, relative to some model, behaves nonoptimally, this must be due to such causes. If this is so, then the animal has features that prevent optimal behavior. But according to the definition of A-constraint, this implies that the strategy set of our model has been defined too broadly, since the animal is A-constrained not to perform the optimal strategy. This line of reasoning applies to all the nonoptimal behaviors that we come across. But if we exclude from our strategy sets optimal behaviors that an animal fails to perform, then we will always end up with a strategy set that makes the observed behavior optimal. Thus, nonoptimal behavior could not exist, for conceptual reasons (Lewontin 1987: 159). This is an unpalatable result if our aim is to develop a general theory of OFT by way of specific models. The result is acceptable, though, if we regard OFT as merely a collection of specific models that chart constraints for particular situations.

Stranger still is the result that we get if we focus on G-constraints instead of A-constraints. If we were to incorporate all the G-constraints in a model, we would end up with a strategy set that contains just one behavior, viz. the behavior that we actually observe! But let us stick to A-constraints. (Henceforth, I shall use the term "constraint" for "A-constraint" where no misunderstanding is possible.)

To the extent that our model building aims at a general theory, the notion of constraint as defined here is too broad. Either we must come up with a narrower notion of constraint, or we must stipulate that OFT models should only cover subsets of constraints on behavior. As far as I know, a more appropriate notion of constraint is not available in the literature. Thus, we have to elaborate subsets of constraints that are appropriate for OFT. Or, correlatively, we must find criteria that result in acceptable strategy sets.

It is hard to explicate the notion of possible behavior. Why has this problem not been recognized in the literature? I conjecture that this is explained by the mundane consideration that it is easy to make the distinction of possible and impossible behaviors for human beings in daily life, which suggests that it should be easy to make sense of it also in science. We know that, while it is impossible for us to do certain things, ample possibilities exist for free choices. On the face of it, if choices were causally determined, options chosen would not be real options since we could not choose otherwise. The notion of free choice apparently does not sit well with causal determination, since our choices do not feel like causally determined processes.

Should we be concerned, then, with the extraneous subject of free will in the analysis of OFT? Up to a point, we should indeed. Philosophy should remind us that the trouble caused in OFT by the possibility–impossibility distinction resembles problems with the explication of free will. Would causal determination be compatible with free will in human beings? Are we free to choose how we act, or is freedom an illusion since all our acts are causally determined? I shall not review here the complex and protracted philosophical disputes about this problem. Instead, I simply take sides by assuming that a position known as *compatibilism* is reasonable. This position amounts to the thesis that causal determination is compatible with free will. Forced acts, not uncaused acts, are the opposite of free acts. Indeed, the very notion of an uncaused act is unintelligible. When we say that we freely choose to act in some way, we mean that we follow our own motives and inclinations. These motives and inclinations are causes at the bottom of the act. For the sake of argument I am simplifying. Motives and inclinations can be compulsory, as in pathological obsession. We do not commonly regard obsessed persons as free. But such details do not matter here.

If this use of the notion of free will is appropriate, then we can make sense of the distinction between possible and impossible behaviors, or acts, in *Homo sapiens*. We cannot choose to fly, because our physical make-up does not allow

The Chimera of Optimality 45

it. But we can choose either to write another paper or to take a holiday. Whatever we do is caused by our mental internal states and dispositions (together with other factors). The set of possible behaviors comprises behaviors not precluded by *other*, nonmental kinds of causes. Does it make sense to say that, if we chose to take a holiday, it would also have been possible for us to write a paper instead? Yes, it does, provided that we mean by this that we would have made a different choice if our mental internal states and dispositions had been different. The implication is not that we are able to divorce ourselves from whatever causal web determines our behavior.

The free-will example indicates how we might arrive at appropriate subsets of constraints for OFT models. We might exclude constraints due to mental internal states and dispositions from the overall set of constraints. That would avert the threat that behaviors become optimal for mere conceptual reasons.

This view of the problem would call for a drastic revision of OFT amounting to a mentalistic approach in line with cognitive ethology. I assume that researchers concerned with optimal foraging would not welcome such a revision. But if we are unwilling to make OFT overtly cognitive, then we should do away with the covert cognitive component through nonmentalistic explications. So let us try to explicate, without recourse to mentalese, the distinction between possible and impossible behaviors.

4.4 HARD AND SOFT CONSTRAINTS

The most promising approach of constraints in OFT I have been able to elaborate is based on the following line of reasoning. Let B_i be the actual foraging behavior or range of foraging behaviors performed by an organism. Then we can define possible, nonactual, behaviors B_j as the set of behaviors generated from B_i by any transformation that could be achieved through causal influences commonly operative in nature.

The idea behind the definition is simple. Shrews cannot fly. No causal scenario is feasible that would change their actual foraging into foraging up in the air. But less drastic changes are feasible.

In line with this, we could distinguish hard and soft constraints on adaptation. Hard constraints cannot be overcome under any causal scenario, while soft constraints can be overcome. OFT models should only incorporate hard constraints if they are intended to support a general theory. The incorporation of all constraints on behavior would land us in trouble, as my analysis has shown. If constraints are restricted to hard constraints, some or even many behaviors may turn out not to be optimal.

Considering learning, for example, we could distinguish between complex foraging behaviors that an animal cannot learn because its neurobiological equipment lacks the sophistication required for that, and foraging behaviors that it simply has not learned as yet because it is facing a new situation but it may learn in the future. The two situations would represent a hard and a soft constraint, respectively. We could describe the former situation by stipulating that foraging is nonoptimal relative to some model due to limited learning abilities, but we could also include constraints on learning in a model and stipulate that foraging is optimal relative to that model. The two descriptions would be empirically equivalent, so optimality versus nonoptimality should be a nonissue in this case. The point would be to describe the foraging behavior and, in part, also to explain it. In either case, we would partially explain it with reference to limited learning abilities as one of the salient causes. In the latter situation, we may be less inclined to model things not yet learned as a constraint, though we can surely do so if we wish. The point in this case would again be to identify a cause of behavior that explains nonoptimality.

Considering natural selection, we could make a similar distinction. Recent changes in the environment may cause animals to forage in nonoptimal ways. This may amount to a constraint on selection since it would need more time to generate optimal behavior. The constraint would be a soft one normally omitted from models. But situations also exist in which selection is unable to generate optimality. If so, then we may decide to incorporate constraints on selection in our models, thus ensuring optimality by model revision. For example, correlation among characters subject to selection may prevent populations from reaching the nearest adaptive peak, as Björklund (1996) and others have demonstrated. Björklund describes the ensuing pattern as nonoptimal. But we could also decide to incorporate this kind of hard constraint on selection in our models, and describe the pattern as optimal. The two descriptions are again empirically equivalent.

The distinction of hard and soft constraints on selection is to some extent arbitrary. It is possible, for example, that a genetic correlation among characters, initially a hard constraint, is ultimately overcome through reorganization of the genome.

Yet another way to distinguish hard and soft constraints on selection would be to define hard constraints as shared constraints—that is, constraints that are invariant across individuals—and soft constraints as constraints that allow of variation in features across individuals. The rationale of this would be that hard constraints thus conceived could not be overcome through selection, while soft constraints could, in principle, be overcome.

The distinction of hard and soft constraints is to some extent arbitrary. We will presumably make different distinctions in different theoretical contexts. As the following example illustrates, other elements of OFT models are likewise arbitrary to some extent.

Consider great tits foraging on caterpillars. A typical prey-selection model will chart relative profitabilities of prey items (say, caterpillars of different sizes), and it will predict which caterpillars the tits should catch and eat. Thus, small caterpillars may be unprofitable because the amount of energy gained from them per time unit is small. The model may be tested in experiments on tits and caterpillars, in the laboratory or in the field. In the field, which should ultimately interest us, a huge quantity and variety of items will be present on which the tits "could" also forage. Let us assume that we observe that they eat caterpillars only. Could they do better if they were also to eat beetles, or flies, or whatever? That question obviously cannot be answered through modeling, because the set of possibilities is huge and diverse. We will typically *assume*, quite possibly on reasonable grounds, that caterpillars are a good choice because they are abundant and easy to catch. Eminently reasonable would be the assumption that stones, or other inorganic materials, would not be a good choice (but I have been told that tits do swallow stones at times).

The strategy set in this case is restricted to possible ways to forage on caterpillars. Notice that the strategy set does not comprise all the behaviors that are possible under whatever hard constraints *on adaptation* we may wish to distinguish. If we exclude eating stones from the strategy set, we in addition rely, implicitly so, on constraints *on maladaptation*.

By way of an alternative, consider a broader strategy set that distinguishes between two kinds of prey: living organisms and other items such as stones. Let our currency be the proportion of living organisms in the diet, and our hypothesis be that this proportion is maximized. The optimal prey choice then would be always to eat organisms and never to eat other items. Relative to this strategy set, the tits would perform very well. But they may not perform well relative to a prey-selection model restricted to caterpillars, with energy intake per time unit as the currency to be maximized.

In practice, we would not model foraging on organisms versus other entities because we already know that foraging in tits is adaptive ("optimal") in that they do not normally feed on items from which they could not extract energy. Moreover, adequate information about this can been obtained by simple observation; no modeling is required. All this is so obvious that we tend to overlook it. Yet the fact that animals do not normally eat entirely inappropriate things is an important adaptation.

Strategy sets in OFT models are always restricted to a subset of possible behaviors, one which is to some extent arbitrary (Lewontin 1987). This may result in a biased view of adaptation. We begin by forgetting about aspects of behavior that we already know to be adaptive or even optimal relative to some standard. Then we model other aspects of behavior to determine degrees of adaptation. If the observed value of our currency variable deviates from the optimal value predicted by the model, then, unless we reject the model for some reason, we conclude that the behavior observed is nonoptimal. That is legitimate, but we should realize that we are dealing only with nonoptimality relative to the strategy set chosen. Optimality (or nonoptimality) in general is not the issue, since it is a meaningless notion. All behaviors are optimal in some respects and nonoptimal in other respects. Our aim should be to find out in *which* respects behaviors are optimal.

From this point of view, optimality models are always useful, even when our tests generate observations that do not fit well with the optimal value of the currency. The idea that we must reject (or modify) a model in the face of failed predictions assumes that the model has the purpose of testing the hypothesis that foraging is optimal relative to the strategy set. But hypothesis-testing is not the only purpose that OFT models can serve. I would regard as more important the purpose of estimating how observed currency values relate to optimal ones. A good fit is informative. So is a lack of good fit. In either case we get useful information about degrees of adaptation *relative to some standard*, and in either case it makes sense to search subsequently for an explanation of the observed behavior.

Those who put hypothesis-testing at centre stage in modeling often assume that research proceeds as follows. (i) We test a particular model. (ii) If observations generate currency values that are nonoptimal according to the model, then it is reasonable to search for additional constraints. (iii) Inclusion of the constraints in the model may give better results. (iv) If not, then we can aim at further revisions until satisfactory results are obtained. Proponents of OFT regard this as a reasonable procedure; opponents argue that it is a reason to reject OFT because the hypothesis of optimality becomes untestable in this way.

Both stances are problematic to the extent that "the" hypothesis of optimality, or OFT as a general theory, is the target of the disputes. On a broad definition of "constraint," the thesis that all foraging behaviors are optimal is true for conceptual reasons. Hence it makes no sense to test this thesis. The critics are right that the thesis is untestable, but this is not a cause for concern. The thesis of optimality should not be the point of models if we aim to include all constraints. Also, that models are corrected in the face of adverse evidence should not be an issue. Scientists do this all the time; otherwise, science would soon come to a grinding halt. The fact that the general thesis of optimality becomes untestable if constraints are exhaustively specified does not imply that models become untestable. The models indicate *in what respects* behavior is optimal in particular cases. That remains an empirical matter decided by tests. Proponents of OFT are right that it may be appropriate to correct "failed" models by the inclusion of additional constraints, but they have not always noticed that this makes general claims of optimality vacuous.

As I have argued, we get a different situation if constraints in OFT models are restricted to hard constraints. In that case, we may end up with a general empirical theory showing that behaviors are always or mostly optimal relative to these constraints. But it is also possible that we will end up with nonoptimality in many cases. That would not militate against OFT, provided that we dispense with general claims about optimality. In this case as well, the remaining purpose of OFT would be to chart in what respects diverse behaviors are optimal or nonoptimal.

The arbitrariness of the distinction of hard and soft constraints is not a problem if researchers are explicit about the distinctions they make in their research. We need some such distinction if we aim to use our models for general empirical

The Chimera of Optimality

claims about optimality. Apart from this, if we do not restrict the set of constraints used in models, we run the risk that the search for optimality becomes a never-ending quest. Restricting modeling to some set of hard constraints amounts to a decision to stop refining models at some point, and to accept that this may leave us with a mixture of optimal and nonoptimal behaviors.

Most researchers would grant that some behaviors are nonoptimal. OFT models *cum* empirical data are likely to reveal nonoptimality under particular conditions. My point has been that we should expect this for *conceptual* reasons that have not been addressed by model builders. Models presuppose that key concepts are perspicuous. In OFT they seldom are. Efforts to theorize about behaviors should allot a more central role to conceptual matters. Without that, sophisticated mathematics applied to foraging behavior will easily lead us astray.

4.5 CONTROVERSIES OVER CONSTRAINTS

Let us have a closer look at the controversy over testability and constraints in optimal foraging research. The issue has been discussed in many publications on OFT and optimality theory in general. I consider a sample that conveys characteristic styles of reasoning.

Schoener (1987), in an overview, maintains that OFT models have been reasonably successful, because observations often match model predictions. In reviewing criticism, he notes that Gould and Lewontin (1979) have argued that constraints, particularly genetic ones, often prevent attainment of optimal states. Schoener grants that we do not know much about the genetics of foraging. But foraging behavior is typically highly flexible, so that genetics might not be worth worrying about. (The genetic constraints discussed by Björklund 1996 mentioned earlier do not concern foraging; Oster and Wilson 1978, unlike Schoener, argue that genetics may constrain foraging.) Gould and Lewontin further claim that adaptationists admit the existence of constraints but are loath to use them in explanations. Schoener argues that this criticism does not apply to OFT, because the inclusion of constraints in models is now common practice. Genetic constraints are not included, but that need not bother us. Other constraints—behavioral, morphological, or physiological ones—are an integral part of many models.

Schoener suggests that if we discover a constraint on foraging in a particular case, we should include it in our model; he does not distinguish between hard and soft constraints. But according to my analysis, this would imply that foraging becomes optimal for conceptual reasons if all existing constraints are accommodated. Failed predictions then merely indicate that we have not uncovered all the constraints, or that the model contains false assumptions. What can tests teach us then? The issue cannot be whether we must accept or reject OFT, given some success rate of testing. If a model is adequately confirmed, we can conclude that we identified relevant constraints. If not, then constraints remaining to be identified should help us replace the model by a better one. In the end, we may obtain much useful knowledge about constraints operative in different cases. That would amount to valuable natural history.

Schoener instead argues that OFT has been verified with tests often and that it should be pursued further for that reason. This suggests that the issue of tests concerns OFT as a general theory, and that refinement of failed models could result in a well-confirmed overall theory.

This view is problematic. Suppose that we will ultimately be successful in the sense that we end up with a collection of models that in each case cover all operative constraints and correctly predict optimal behaviors. What would that imply? The conclusion should not be that OFT in general would be well confirmed, because OFT would be true for conceptual reasons. Instead, we would have diverse models showing that optimality can take many different forms. That would amount to natural history, not theory in a strict sense.

Gray (1987), in an overview in the same volume, criticises OFT in harsh terms. In his view, OFT has not done well under tests. True, some model assumptions have been confirmed, but other assumptions are seldom met. I will not try here to sort out the causes of the discrepancy between the evaluations by Schoener and by Gray. They agree, anyhow, that some positive results have been obtained. As I explained, negative results cannot militate against OFT as is commonly construed. Gray's view of tests, for that matter, is inconsistent because he also argues that OFT is not testable. If that were right, then (dis)-confirmation should be impossible. It is strange that many authors, like Gray, have argued that OFT is untestable *and* false (Pierce and Ollason 1987 provide another example).

Gray has many other points of criticism. I focus on what he says about constraints:

The final major assumption of OFT is that natural selection is the only force that has moulded variation in evolutionary time. [But that is demonstrably false.] ... However, incorporating ... nonselective phenomena in anything other than a trivial manner would undermine OFT as a general approach to foraging behavior because there is no way of distinguishing between the effects of structural and historical constraints, the direct action of the environment, molecular drive, evolutionary lag, and chance in shaping the current diversity we see. OFT highlights the fact that functional explanations force researchers to be panselectionist in practice even if they are not in theory. [Gray 1987: 74]

Gray assumes that many different constraints on foraging are operative in nature, and that we are unable in practice adequately to model them all. Thus we are forced to disregard constraints, and that amounts to panselectionism. This criticism is unfair, because OFT models do contain *some* constraints. If some models generate correct predictions, they confirm the assumption that, in some cases at least, foraging is restricted only by a limited number of constraints as modeled. Gray may be right that, in other cases, the sheer complexity generated by diverse

The Chimera of Optimality

constraints prevents the elaboration of a model that incorporates them all. That would only mean that we cannot always elaborate models that show, comprehensively so, *in what respects* observed behaviors are optimal

Like Schoener, in their well-known book Stephens and Krebs (1986) regard OFT as reasonably well confirmed. They note that first-generation models, which are not always successful, postulate few constraints. Second-generation models have improved predictions by the inclusion of more constraints. It is reasonable to refine models by adding constraints in the face of adverse evidence. This need not amount to ad-hocism, as new constraint assumptions can be tested.

Would it be possible also to test OFT as a general theory? Stephens and Krebs dismiss this question. *The* optimal way to forage does not exist. Hence OFT is not one general theory. "It is a point of view, comprising a large and continuously changing set of models" (p. 199).

Stephens and Krebs do not appear to think that we must restrict the set of constraints that qualify for inclusion in models. I wonder if they would accept my thesis that any behavior must be optimal for conceptual reasons relative to an exhaustive set of constraints. The statement that "ultimately, foraging theory must account for observed foraging behavior" (p. 205) suggests that they might agree with this in principle. It is desirable that researchers be more explicit about this issue.

Parker and Maynard Smith (1990) state that, according to optimality theory, adaptation is pervasive but not perfect. The aim of the theory is to understand behavior in terms of selective forces and constraints. If observations do not match the predictions of a particular model, we need to rework model assumptions and retest the improved model. "This has been criticised as being an iterative procedure leading inevitably to a fit. But this is how science works; theories can only be discarded when they are disproven or found to be unrealistic" (p. 29).

This remark suggests that Parker and Maynard Smith grant that good modeling leads inevitably to a fit, so that a general optimality theory must be true for nonempirical reasons. The emphasis should, rather, be on specific forms of optimality as portrayed by successful models. But other remarks suggest that the authors have a different reconstruction in mind. They note that most optimality models assume that phenotypes breed true, that pleiotropy does not operate, that strategies replicate independently of each other, and so forth. "Obviously, selection cannot produce an optimum if there is no way of achieving it genetically, but for some models, it is clear that selection will get as close to the optimum as the genetic mechanism will allow" (p. 31). This suggests that not all constraints need to be accommodated in models. Hence we get nonoptimal behaviors, but we can identify the constraints responsible for nonoptimality.

The two reconstructions are, in fact, equivalent. If we choose to have all the constraints in our models, then we always get optimal behaviors for conceptual reasons, and the constraints modeled will tell us what form optimality takes in different cases. If we do not accommodate all constraints, then some behaviors will be nonoptimal for empirical reasons, and constraints outside the models will explain nonoptimality. At first sight, the two reconstructions are contradictory ("All behaviors are optimal" versus "Not all behaviors are optimal"), but they are actually compatible since they involve different construals of optimality. In any case, I assume that Parker and Maynard Smith would agree that, to a large extent, optimality modeling amounts to doing natural history.

The authors I mentioned share an emphasis on specific models in tests of OFT. Other authors apparently envisage OFT primarily as a general theory. For example, I assume that the harsh criticism of optimality theory voiced by Lewontin (1987: 159) must be seen in this light. Here is how he sums up his analysis:

> The purpose of the argument is not to refute optimality theory but to demand its nontrivial formulation. If the principle of optimality is to be useful as a description of what actually happens in evolution, it cannot swallow all accidents of history. If random genetic drift ... is really at all common, then optimisation is *not* a good characterisation of the evolutionary process. The optimisation claim must be ambitious enough to exclude a good deal of accident or it becomes empirically vacuous. It must, however, not be so ambitious as to exclude all historical contingency since then we would know it to be untrue a priori. Between these two it is not clear to me how much space is left for enlightenment.

Lewontin is concerned with "the" principle of optimality. So he would presumably regard OFT as a general theory. One horn of the dilemma that Lewontin sees concerns the empirical vacuity we get if we do not "exclude a good deal of accident." In my terminology, the problem is that OFT becomes true for conceptual reasons in the absence of limits on constraints. I would add that this would leave us with valuable natural history. Lewontin does not recognize this, as he is concerned with general theory.

The other horn of the dilemma is that the exclusion of much contingency would leave us with a false theory. In other words, severe limits on constraints allowed in models would imply that much behavior is not optimal. Again, we would be left with valuable natural history, not with acceptable general theory.

Lewontin calls, in my terminology, for a distinction of hard and soft constraints. He notes that no adequate distinction has been made by foraging theorists. Furthermore, he suggests that inclusion of soft constraints in models generates a vacuous theory, whereas exclusion of these constraints leaves us with a false theory. My conclusion, instead, would be that the first option would provide us with valuable natural history, admittedly not general theory. The second option might result in a general theory to the effect that foraging is always or mostly optimal relative to hard constraints. If such a theory would turn out to be false, as Lewontin suggests, we would have to resort to more specific claims about optimality and nonoptimality in particular situations. That would again represent a move in the direction of natural history.

Orzack and Sober (1994) also construe OFT as a general theory, and they are explicit about the distinction needed. They note that optimality analysis focuses

The Chimera of Optimality

on the determination of evolutionary stable phenotypes. According to them, a strong thesis of optimality, which they also call strong "adaptationism," is best construed as follows. "Natural selection is a sufficient explanation for most nonmolecular traits, and these traits are locally optimal" (p. 364). Global constraints (roughly my "hard" constraints) are compatible with this; optimality theorists take such constraints for granted (see also Sober 1995–1996). But adaptationism demands that local constraints (roughly my "soft" constraints), which are linked up with variable traits, do not restrict optimal behavior. This implies, according to adaptationism as construed by Orzack and Sober, that natural selection should be overridingly important causally in the generation of phenotypes, whereas other evolutionary forces such as mutation pressure and genetic drift should have a negligible role.

Orzack and Sober's view of the influence of natural selection is somewhat arbitrary. In their view, the thesis of optimality assumes that selection is a sufficient explanation for most nonmolecular traits. Selection is presumed to work under global (hard) constraints represented by invariant features. Now, such features may well be the result of mutation pressure or genetic drift in the past. If so, then they would nonetheless not militate against the thesis of optimality, because they are now invariant. But visible effects of mutation pressure or drift in the present would count against optimality. This view has counterintuitive implications. Thus, an evolutionary process in which a feature goes to fixation and so becomes invariant due to drift should be construed as a transition from a nonoptimal state to an optimal state, even if the feature would be maladaptive compared to alternatives eliminated from the population.

It is conceivable but improbable that Orzack and Sober's version of optimality theory will be confirmed in the future. We know that natural selection is not the only force responsible for evolution. No reason exists to assume that other forces are much less important. Disconfirmation of their version of the theory would not imply that we must abandon optimality modeling. A natural option in the face of disconfirmation would be to enlarge the set of constraints allowed in models. Orzack and Sober have not, in any case, provided us with a nonarbitrary set of allowable constraints.

I have argued that, if we do not limit allowable constraints and strategy sets, any behavior conceivable will be optimal relative to some specific model. Thus all behaviors would be optimal in some sense for conceptual reasons. This result is blocked by my suggestion that constraints be limited to hard constraints. My proposal for a general core of OFT indicates that further limitations are desirable. Constraints and strategy sets generating "optimal" behaviors that are utterly maladaptive should not be allowed. In line with this, all OFT models presuppose constraints on maladaptation which, unlike constraints on adaptation, remain implicit. Thus a strategy set portraying ways to forage on live prey presupposes that animals are constrained not to forage on stones. We do not make such things explicit because they are obvious. However, if we forget about the obvious we get a distorted picture of OFT.

If we accommodated all constraints on behavior in OFT models, the ensuing optimality, which would prevail for conceptual reasons, would not confirm the general core of OFT. But the models would still constitute a test in a different way. Suppose that it turned out that, to get optimality, we have to include a great variety of constraints that would reduce the strategy set to one behavior: the behavior observed. That would mean that behaviors are maladaptive to an extent that flies in the face of the general core of OFT. We would have to reject OFT in this case. OFT obviously presupposes that models with constraints that ensure optimality should have strategy sets comprising possible behaviors that deviate from actual behaviors. The fact that many models do have this feature confirms the general core of OFT in a qualitative way.

4.6 CONCLUSIONS

My analysis shows that OFT has a general core theory. I characterized this theory in succinct qualitative terms. It is easy to show that the theory, if construed in this way, is testable. Indeed, it is well confirmed, by evidence that is mostly disregarded in the literature. Researchers have overlooked the obvious since their reconstructions of the core theory rely on a general notion of optimality which is meaningless.

Concerning specific models, the situation is less clear. An important point is whether such models warrant universal empirical claims. At present there are no well-confirmed claims of this kind. This need not imply, though, that specific models are useless. Possibilities for universal claims are limited in biology. Instead we must often be content with natural history. I regard natural history as an immensely valuable part of biology.

Whenever a behavior does not conform to the optimal outcome generated by a model, it is reasonable to postulate additional constraints with the implication that the behavior is optimal after all. The general thesis of optimality thus becomes true by definition. The point of OFT is obviously not to show that optimality is widespread. Its purpose is, rather, to show what types of constraint are operative in different species and in different situations.

If we aim to include all constraints on behavior in our models, we will always end up with optimality. But we may also choose to distinguish constraints that cannot be overcome (hard constraints) from constraints that can be overcome in principle (soft constraints), and to restrict models to constraints in the former category. Thus we may repeatedly come across nonoptimal behaviors. This does not imply that we are dealing with useless models. Nonoptimality would be a result we can explain by the identification of soft constraints.

5

The Units of Selection

5.1 INTRODUCTION

Evolution involves genetic change in populations. Understanding typical examples of evolution amounts to knowing how natural selection effects this genetic change. Selection involves differential reproductive survival of phenotypes in populations. Populations show variation in phenotypes. If phenotypes differ in reproductive survival, and if the relevant phenotype features are heritable, then changes in the relative frequencies of the phenotypes ensue. Ultimately some phenotypes may go extinct while others survive. Likewise for genotypes.

From the most common point of view, organisms are the primary "units of selection." Consider the well-known example of industrial melanism. In areas polluted by industry, the trunks and branches of trees become dark by the deposition of soot. If moths with a light color stay on these trees, they are conspicuous against the background and are easily spotted by predators. Mutants with a dark color are thus at an advantage. Selective predation pressure causes light morphs to decrease in frequency and dark morphs to increase in frequency. At the same time, the frequencies in underlying genes will change. But selection by predation does not directly act on genes. Predators do not eat moth genes; they eat moths. Hence the idea that organisms are the "unit of selection."

Conceivably, natural selection acts not only on individual organisms, but also on higher-level units such as groups and species. Disputes over the "true" level of selection tend to be confused as different researchers use different criteria to distinguish selection at different levels (for references and comments, see for example Brandon 1990; Sober and Wilson 1994; Wilson and Sober 1994).

Genes are replicators, whereas individuals are vehicles (terminology of Dawkins 1982) or interactors (terminology of Hull 1980). The vehicles interact with the environment in such a way that replication ensues. The dispute over levels of selection concerns the nature of vehicle selection.

Considering units of selection, we should distinguish the controversy over vehicle selection from a different controversy, which concerns the role of genes in processes of selection. Some allot a central role to genes. Others argue that individuals (or groups, or species, as the case may be) should be at centre stage. Genic selectionists would not deny that individuals (and possibly groups, species) play roles in selection processes, but they prefer to focus on genes in models and theories of evolution. They choose to regard replicator selection as more important than vehicle selection.

I argue in the next section that the controversy over genic selectionism has no substance. It merely concerns differences in preferred modes of presentation, the value of which depends on the purposes of research (for additional comments, see also Hull 1988c: 412–414). Subsequently, I criticize the view of Brandon (1990), who appeals to the notion of screening-off to explain why organisms rather than genes are the prime target of selection. The rest of the chapter analyzes controversies over vehicles which have more substance, although they suffer from much terminological confusion.

The analyses in this chapter confirm the view that evolutionary theory consists of natural history rather than laws of nature. The equations of population genetics are perhaps the most obvious candidate laws. But the equations are meaningless without a context of interpretation, and interpreted equations fail to function as laws.

5.2 GENIC SELECTIONISM: THE ISSUE OF REPRESENTATION

The dispute over genic selectionism is odd because the disputants do not disagree about facts. The dispute appears to concern philosophical interpretations of factual issues. Genic selectionists, as we understand them, regard genes as causally more important than other entities. Thus, they would opt for a representation of selection processes in genic (allelic) terms. Opponents opt for different representations of some selection processes, which allegedly capture causation in more appropriate ways. Causation is a pivotal issue in the dispute (e.g., see Brandon 1990; Godfrey-Smith and Lewontin 1993; Sober 1984).

By way of an example, I analyze one of the most incisive criticisms of genic selectionism, by Godfrey-Smith and Lewontin (1993). These authors argue by an analysis of examples from population genetics that genic representations of selection in equations of population genetics, however accurate, do not properly represent the causation of all selection processes. Some selection processes, they propose, are best represented by genotypic models—that is, models dealing with genotypes at the locus considered. Other opponents of genic selectionism concentrate on phenotypes, not genotypes (e.g., see Brandon 1990). For ease of exposition, I only consider genotypes. *Mutatis mutandis*, my arguments would also apply to analyses accounting for phenotypes.

Godfrey-Smith and Lewontin regard as crucial the number of dimensions (in the sense of axes for variables that define a model space) needed to model particular selection processes. They further characterize models by rules of transformation describing how the system modeled moves in state space, and by parameter values. The examples of selection they consider involve one locus with two alleles.

The notion of cause is pivotal in Godfrey-Smith and Lewontin's arguments. That calls for a caveat. The notion is notoriously problematic. Disputants concerned with genic selectionism depart from different interpretations of "cause" (e.g., Brandon 1990; Sober 1984). No wonder then, that opposing parties do not manage to terminate the disputes. I do not take sides in this thorny issue. For the sake of argument, I take over the notion of cause as an undefined item. Later I show that the emphasis on causation is misguided, regardless of the precise meaning of "cause."

Heterozygote inferiority is the first example that Godfrey-Smith and Lewontin consider. In brief, they reason as follows. Suppose, for some two-allele locus, that the two homozygotes are equally fit (relative fitness = 1), while the fitness of the heterozygote is lower $(1 - s)$. Under the assumption of random mating, we can compute relative frequencies of mating types from the original genotype frequencies. Genotype frequencies for the next generation are then inferred from the old frequencies. The new genotype frequencies can also be expressed as a function of the old allele frequencies, and trajectories for allele frequencies can be obtained that do not refer to genotype frequencies. Thus, if p and q are the original relative frequencies of two alleles A and a, the change in the frequency of A in a generation, Δp, is $pqs(2p - 1)/(1 - 2pqs)$.

A one-dimensional allelic space can represent the evolutionary process in this case. Yet, heterozygote inferiority is an example of selection at the level of individuals with particular genotypes, since the parameter s measures the systematic loss of individuals. Godfrey-Smith and Lewontin argue that a genic representation therefore does not properly capture causation.

Next, Godfrey-Smith and Lewontin consider the example of Rh blood groups in man. Here, genotypes are assumed to have equal fitnesses, with one exception. A heterozygote foetus in a homozygote rhesus-negative mother has a lowered fitness due to antibodies against the foetus produced by the mother. In this case, a two-dimensional genotypic model will aptly represent the selection process. However, it is also possible to model the selection process with a two-dimensional allelic space, by using conditional allele frequencies (frequencies of alleles in a genotypic context). So model representations as such, in this case also, do not allow inferences about causation. It is, rather, the other way round. We must know about causation to decide what representation of a particular selection process is appropriate.

This also holds for the next example, discussed by Godfrey-Smith and Lewontin—gametic selection where sperm containing different alleles have different chances of fertilizing an egg cell due to differences in swimming speed. This amounts to genic selection, but a two-dimensional model is needed here to represent the selection process. Dimensionality in itself obviously does not suffice to distinguish between genic selection and other forms of selection.

Godfrey-Smith and Lewontin's main point is negative: model representations of selection do not allow inferences about causation. They admittedly are unable to present a hard-and-fast method to identify patterns of causation. One of their suggestions is that we can examine the role of model parameters. Thus, the parameter s in the heterozygote-inferiority example points to a process of selection in which individuals with a particular genotype have a central role.

In the terminology of replicators and vehicles, which Godfrey-Smith and Lewontin do not use, haploid sex cells are the vehicles in gametic selection. That is why the genic rather than the genotypic perspective is appropriate. In the other examples, diploid individual organisms are the vehicles, so that the genotypic perspective is more appropriate. Genic selectionists are unlikely to be impressed by this line of reasoning. They may simply state that vehicles are not their concern and that selection is primarily replicator selection.

Considering equations of population genetics, the parties in the dispute over genic selectionism agree that genic representations of selection processes are always possible. But Godfrey-Smith and Lewontin regard some of such representations as inadequate because they do not fit in with the causal processes involved.

My response here would be that, to come to grips with causation, in any appropriate sense, mere equations without context will not suffice. Furthermore, I would argue that, once the context is made explicit, reference to processes involving genotypes in addition to processes involving genes is unavoidable. If so, then the nature of the disagreement becomes elusive.

Without a proper context of explication, an equation such as

$$\Delta p = pqs(2p - 1)/(1 - 2pqs)$$

does not provide information about the process modeled. To provide information, the equation must be expanded such that the intended meaning is clear. I would propose replacing it by the following statement:

> For any locus with two alleles, A and a, with frequencies p and q, if the fitness of AA and aa is 1, and the fitness of Aa is $(1 - s)$, and conditions such as random mating and absence of mutation and migration are satisfied, then the change in frequency of A in one generation equals $pqs(2p - 1)/(1 - 2pqs)$.

The Units of Selection

This statement captures the intended meaning of the equation. It makes explicit the empirical content that the formula as such lacks. Now the statement does not tell the entire story of heterozygote inferiority. The entire story—that is, the full-fledged theory of heterozygote inferiority—tells us how to derive this statement from premises about the frequencies of genotypes of offspring deriving from various matings. Godfrey-Smith and Lewontin do tell full stories. But their discussion of genic selectionism is restricted to the equations of population genetics. Now, without a context of explication that provides for the meaning of variables and parameters, equations are meaningless. An equation for selection against heterozygotes does not explicitly refer to genotypes. But it covers genotypes in an implicit way. This is also true, we should add, for equations in the writings of genic selectionists. Neither party in the dispute over genic selectionism would deny that, once the context presupposed by equations is made explicit, reference to processes involving genotypes is unavoidable.

The point of the controversy appears to be that full stories can be told in different ways. We need equations and a context. In modeling selection, we cover some processes with equations, while relegating the description of other processes to the context. This amounts to a partitioning of information over equations and context.

Genic selectionists and their opponents resort to different partitionings in presentations of selection processes. They should agree that both the equations and the context are needed for a proper understanding of selection processes. The disputants would draw the line between equations and context in different ways. That is their privilege. Nothing much follows from this regarding fundamental roles for particular types of entity (gene, genotype, and so forth). Line drawing is a pragmatic matter.

A proper theory of evolution due to heterozygote inferiority will deal with many processes (mating, reproduction, death) involving individuals with different genotypes, and it will thereby chart changes in genotype frequencies and gene frequencies. If the combination of these processes is termed "the process of natural selection against heterozygotes," then the full story about selection involves individual organisms, genotypes, and genes.

Godfrey-Smith and Lewontin might regard my analysis as unfair because they do grant that genic representations implicitly refer to genotypes. Hence, the mere fact that genic selectionists privilege equations with variables for allele frequencies should not in itself militate against them. Godfrey-Smith and Lewontin's salient point is rather that genic representations do not faithfully represent causation. I shall show that this appeal to causation does not work.

Considering dynamic sufficiency, Godfrey-Smith and Lewontin explicitly grant that no privileged state space exists. Genic equations and genotypic equations may be equally adequate in this respect. But they hold that privileged causal stories exist. The selection processes analyzed by them involve different patterns of causation. Models should distinguish between these patterns. They opt for models that best represents causal relationships.

This line of reasoning is misguided in that mathematical equations without interpretation are an improper means to express causation. Philosophers disagree over the explication of the notion of cause, but they would agree that mathematics as used in population genetics is not sufficiently rich to express causation. Hence, faithful representations of causation should draw on equations of population genetics together with some context of interpretation.

My comments concerning partitioning remain in force here. Genic selectionists and their opponents partition information about selection processes in different ways. When they opt for equations that provide different types of overt information, contexts of interpretation will have to compensate for the difference. Full stories of selection cannot be told without contexts for equations.

The issue is whether full stories should deal with causation. Let us assume that this is a reasonable demand. We then get two stories that cover the same subject through different partitionings. Either story may adequately deal with causation. The controversy over genic selectionism thus dissolves, since the opposing parties appear to agree about facts.

Genic selectionists may still argue that, in selection processes, genes are always more important causally than genotypes or phenotypes. I would be at a loss to understand this. It seems to me that judgements concerning importance reflect context-dependent interests of researchers. It is also conceivable that genic selectionists, unlike Godfrey-Smith and Lewontin, do not wish to address causation in their stories. That would be their privilege. The opposing parties would then be talking about different subjects. The controversy dissolves in this case also.

My thesis that evolutionary theory should be regarded as natural history is underlined by the fact that the equations in models of population genetics are meaningless without a context of interpretation. True, interpreted equations are universal claims. But such claims apply only to idealized systems that satisfy many assumptions that are seldom jointly realized in nature. Furthermore, particular phenomena can mostly be explained by different models. Consider, for example, the explanation of variability in the form of polymorphism, with genes represented by different alleles in a population. Models show that this phenomenon can be explained by, for example, heterozygote superiority, some forms of frequency-dependent selection, and some instances of spatial heterogeneity. The models in their turn represent heterogeneous collections of phenomena. This leaves us with the general, qualitative claim that natural selection can generate polymorphisms through a variety of unrelated mechanisms. Neutral mutations, of course, can also generate polymorphisms. In principle, we can find out how common various mechanisms are, but we have no laws to help us with this.

5.3 THE PUZZLE OF SCREENING-OFF

Selection primarily acts on individual organisms, not genes. Brandon (1990) has tried to make this criticism of genic selectionism more precise by appealing to the notion of screening-off. In this section, I analyze his view and present criticism by opponents (Sober 1992), together with a rebuttal (Brandon et al. 1994). The analysis uncovers some tricky problems concerning the conceptualization of selection processes. In a loose manner of speaking, selection amounts to differential survival of phenotypes, which entails differential survival of genotypes. The genotypes produce the next generation of phenotypes, and so forth. This evokes a picture of different, successive processes that distorts reality. I use the dispute about screening-off to unearth this misleading picture.

Screening-off is defined in terms of conditional probabilities: P screens off D from E iff $\Pr(E, P\&D) = \Pr(E, P) \neq \Pr(E, D)$. In words, P screens off D from E iff the probability of E given that $P\&D$ equals the probability of E given that P and does not equal the probability of E given that D. According to Brandon (1990: 83–84), in standard cases of selection, the phenotype screens off the genotype with respect to reproductive success. His example is directional selection for increased height. Here tall organisms on average have a higher fitness than short ones. Since there is genetic variation in height, fitness is also associated with certain genes or genotypes. Yet natural selection does not favor phenotypes and genes equally (there is "asymmetry"), because reproductive success is determined by phenotype irrespective of genotype. Selection "sees" particular heights; it does not also "see" the genes associated with height. This idea is made precise by the screening-off relation. Phenotypes screen off genes from reproductive success. Hence, Brandon says, phenotypes are better causal explainers of reproductive success than are genes. He also notes that phenotypes are proximate causes and genes are remote causes; proximate causes are thought to screen off remote causes from their effects.

Sober's first worry about the asymmetry is that "the description of the phenotype properties . . . must be *complete*; otherwise, there is nothing to prevent a genotypic specification from affecting the organism's prospects for survival even after the phenotypic character is taken into account" (1992: 143). Brandon et al. (1994) reply that descriptions and specifications do not matter: "the *objective* probability of [reproductive success] for *this* organism with *its* phenotype is unaffected by how well or how completely we describe its phenotype" (p. 479).

To Sober's point I would add that we cannot do without descriptions, and that it is impossible for descriptions to be complete. The way Brandon et al. phrase their reply indicates that they are using the expression "phenotype" for the entire set of phenotype features of an organism. Likewise, I presume, for "genotype." Let us suppose, for the sake of argument, that their screening-off relation, interpreted in this way, generally holds. Then there is still a long way to go to concrete evolutionary explanations that refer to particular phenotypes and genotypes. In such explanations, we have to generalize over individuals. Now generalizations are impossible if entire phenotypes and genotypes are our starting point, because phenotypes and genotypes of individuals are unique (unless we are dealing with asexual organisms or identical twins). In explanations, therefore, we must describe and specify phenotype features and/or genotype features. Organisms can share features, not entire phenotypes or genotypes. As a matter of fact, Brandon (1990) couched his example in terms of a particular phenotype feature—height— and particular genes affecting this feature.

To alleviate Sober's worry, Brandon et al. would have to assume that the screening-off relation generally holds for sensible choices of phenotype features and allied genotype features. Let us grant this. (In fact, it is not true: Sober and Wilson 1994 provide examples where the screening-off relation does not hold.) Would it imply that we have laid our hands on proper *causal* explanations linking reproductive success with phenotype (which is what Brandon et al. are after)? I doubt it. We could at best formulate explanations that represent causal processes in an indirect way. The effects of phenotype features will normally be context-dependent. A particular phenotoken feature may work out well in the context of one phenotype and be a handicap in the context of another phenotype. Probabilistic relations between phenotype features and reproductive success are summaries of heterogeneous effects. They do not represent straightforward causal relations.

Brandon et al. state that the issue is not how well or completely we describe the phenotype of an organism. At issue are objective single-case probabilities of reproductive success, which depend on the entire phenotypes of individual organisms. This response of Brandon et al. to Sober's criticism is misleading because it suggests that, in forging explanatory links between phenotypes and reproductive success, evolutionary biologists would like to describe entire phenotypes, which they cannot, so that they have to be content with approximate descriptions that do not cover phenotypes well or completely. I would rather say that evolutionary biologists do not aim to describe entire phenotypes at all. True, they will grant that entire phenotypes are causally relevant, but in their models, theories, and explanations they concentrate on phenotype features and so generalize over sets of diverse phenotypes. That does not amount to an approximation of complete phenotypes; it is an entirely different approach.

Brandon et al. are concerned with ontological relationships that are not covered by explanatory relationships as formulated in evolutionary biology. Their ontology might be appropriate (see, however, section 5.4), but it does not even remotely represent actual explanations. The ontology presented by Brandon et al. is unhelpful in the appraisal of actual evolutionary theory. I argue that the ontology is problematic. My criticism is directed against the views of Sober and Brandon alike.

Brandon (1990: 83–84) holds that proximate causes screen off remote causes from their effects, and that proximate causes are better explainers than remote causes. Both Sober and Brandon use this terminology to describe temporal

The Units of Selection 63

"steps" in selection processes. They appear to assume that genotypes are remote causes of reproductive success whereas phenotypes are proximate causes. I argue that this is a potentially misleading conflation of dichotomies. Genotypes are not temporally prior to phenotypes. Organisms have genotypes and phenotypes throughout their existence (see Oyama 1985 and Griffiths and Gray 1994, who chart consequences of this for evolutionary biology in a different context).

A case in point is Sober's criticism of genic selectionism (1984: 226–233). From a common viewpoint, genes, phenotypes, and reproductive success are links in a causal chain. Genic selectionists would argue that genic explanations are "deeper" because they refer to remote, not proximate causes. Sober (1984: 230) provides an explicit reconstruction of the argument, in which the conclusion that genes cause reproductive success is inferred from three premises: (i) genes cause phenotypes, (ii) phenotypes cause reproductive success, and (iii) causality is transitive. He notes that opponents of genic selectionism have rejected either premise (i), their argument being that ensembles of genes rather than individual genes are causally efficacious, or premise (iii). Sober agrees on both counts: polygenic effects warrant the rejection of premise (i) (p. 313), and premise (iii) is invalid at the population level (pp. 297–298). This line of reasoning does warrant the rejection of genic selectionism. Indeed, Sober offers many more insightful arguments, couched in terms of his distinction of "selection *of*" and "selection *for*," against genic selectionism. However, he appears to share with genic selectionism the assumption that premises (i) and (ii) correctly describe temporal phases of selection processes, provided that we replace "genes" in premise (i) by "genotypes."

Brandon distinguishes similar temporal steps in processes of selection. In one step, phenotypes interact with their environment in a way that causes differential reproduction; in the next step we get differential replication of genes; and a third step—epigenesis—provides the link between genotype and phenotype (Brandon 1990: 79). Both Sober (1992) and Brandon et al. (1994) appear to retain this image of selection.

Let us concentrate on the idea that genotypes cause phenotypes. Do genotypes—together with the environment—indeed cause phenotypes? It seems to me that we are dealing here with a highly elliptical manner of speaking. Zygotes develop into adult organisms, and organisms have a genotype and a phenotype throughout development. The phenotype changes drastically in the process, and the genotype is more stable. However, genotype features continually play causal roles, just as do phenotype features.

Consider selection in temperature adaptation. A process of viability selection that leads to differential mortality in organisms that produce different enzymes regulating metabolism as a function of temperature will continually involve a web of subprocesses in which both genotype features and phenotype features play causal roles. The enzymes result from DNA transcription, and the transcription itself is influenced by all sorts of phenotype features. It would be misleading to say that we are dealing here with processes in which genotypes cause phenotypes. Instead we have continual interactions between genotype features and phenotype features (and features of the environment).

Genotype features and phenotype features causally interact throughout the organism's existence. If we want to say that the genotype (together with the environment) causes the phenotype, then we must somehow partition the causal web such that we get a separate compartment of causal one-way traffic from genotype to phenotype. I would not know how to characterize the one-way traffic in a general way. By implication, it becomes unclear how the phenotype could screen off the genotype from reproductive success.

So much for one "step" in the process of selection—genotypes (together with the environment) causing phenotypes. The other "steps"—phenotypes (together with the environment) causing reproductive success, and reproductive success leading to differential replication of genes—are not real steps either. In viability selection and in fertility selection, entire organisms die or reproduce. If an organism dies, for example, its phenotoken disappears. So does its genotoken. The disappearance of the genotoken is part of a "process" of "replication" (or failure of replication) if deaths are differential; the disappearance of the phenotoken is a different "process," but the two "processes" are actually one process.

Processes of selection do involve a temporal succession of steps—namely, reproduction followed by development of the offspring followed by death. We can get selection if steps are different in different types of organism. Selection leads to changes in genotypes and phenotypes. Both genotype features and phenotype features play causal roles in all the steps. Now, it is hardly feasible to model all the causal roles in a realistic way. We have to simplify. Considering development, we can do that, for example, by charting relations (relative to environments) between genotypes after fertilization and adult phenotypes. If we are lucky, we get clear relations. But we should beware of thinking that we have thereby modeled how the genotype causes the phenotype.

If we were to choose to simplify along the lines portrayed by Brandon and Sober, we would get a picture that is much more complex than they suggest in the discussions about screening-off. Lewontin (1974: 12–16) has presented a nice scheme of relations—in his words, "laws of transformation"—that we would then need fully to cover selection. We need four kinds of laws: epigenetic laws (representing phenotypes developing from genotypes), laws that describe changes in phenotype composition in a population within a generation, laws that allow us to infer genotype composition from phenotype composition, and laws that show how reproduction changes genotype composition. The laws are artificial constructs that do not represent causal processes in a direct way. It is impossible in practice to combine all the constructs into integrative models of population genetics. (Lewontin stated this in 1974; it is still true.) We have to simplify further and disregard some relations in any model that we construct, in the hope that this does not make much difference. But, as I argued, this simplification leads to serious theoretical problems.

I have argued that to get at models and theories we must generalize over individuals, and we cannot do that for entire genotypes and entire phenotypes. So we concentrate on particular genotype features and phenotype features, but that is problematic because of interactions between features.

Let us return to Brandon's view of proximate and remote causes. He argues that phenotypes are better explainers of reproductive success than are genotypes, because phenotypes are proximate causes and genotypes are remote causes, and proximate causes screen off remote causes from their effects. We have just seen that the phenotype/genotype distinction does *not* coincide—in reality, as opposed to some artificial models—with any proximate/remote distinction. Brandon, like Sober, is conflating dichotomies that should be kept apart, in our ontology if not in our models.

The ontology presupposed by both Brandon and Sober fails to illuminate evolutionary theory. To the extent that ontology can be helpful for biology at all, we must search for a different one.

What I said about genotypes and phenotypes makes the common distinction of replicators and vehicles or interactors problematic. Those who use the distinction suggest that replicators (together with the environment) generate vehicles, which are involved in causal interactions influencing the perpetuation of replicators. This again invokes the picture of distinct processes that, in fact, are not distinct at all. It is true, though, that units—misleadingly called vehicles—at different levels of organization may be involved in evolutionary processes. In the rest of this chapter, I analyze this issue.

5.4 GROUP SELECTION AND SPECIES SELECTION

So far, I have been concerned only with selection within populations, with individual organisms characterized by genotypes or phenotypes as vehicles. Would it be possible that selection also operates at higher levels of organization, with groups of organisms within species, or even entire species, as vehicles? For a long time, adherents of the modern synthesis in evolutionary biology have assumed that no separate processes of group selection or species selection exist, and that individual selection should suffice to explain "selection" at higher levels. But in the last few decades, higher-level selection has become popular again.

Considering genic selectionism, I concluded that disputes over it are sterile. Both genes and organisms as vehicles play causal roles in all selection processes. We may choose to focus on genes and relegate organisms to an implicit context of research, or vice versa, but that is a matter of pragmatic interests without ontological import. Would this consideration also apply to individual selection versus group selection? Up to a point, this is indeed so. Causal roles of groups can be redescribed in principle as causal roles of individuals, if group members are conceptualized as part of the environment of individuals (for details and references, see Sterelny 1996). So some selection processes can be described both as group selection and as individual selection. But this situation is not fully analogous to the genic selectionism issue, because all cases of individual selection cannot also be described as examples of group selection. Sometimes groups play causal roles, sometimes they do not. Hence group selection is not merely individual selection by another name, and we should distinguish between the two kinds of selection. But how should we define group selection? That is a moot point, because groups can play causal roles in different ways. It is not surprising, therefore, that different authors subscribe to different definitions. I argue that nothing much hinges on this as long as we distinguish between different causal roles of groups. A single label does not suffice to cover sundry causal roles of groups, and controversies over group selection are mostly a matter of linguistic convention. Likewise for species selection.

Let us start with different meanings of "group selection." Two overarching categories of group selection have been distinguished (Damuth and Heisler 1988; Mayo and Gilinsky 1987). First, group membership could confer fitness values to individuals in groups such that individuals have different fitnesses depending on the group they are in. Let us call this Group Selection (1). Second, group properties or processes could cause differences in fitness among groups in the sense that we get differences in the extinction or production of new groups. Let us call this Group Selection (2).

Group Selection (2) has been demonstrated convincingly in Wade's famous experiments with the flour beetle, *Tribolium castaneum* (McCauley and Wade 1980; Wade 1982; Wade and McCauley 1980). In the experiments, groups of beetles were used to found new groups depending on their productivity (reproductive output). Productivity appeared to be a heritable group feature that could not be entirely explained by selection acting on productivities of individuals. We should notice here that group productivity may nonetheless be a function of other features of individuals and relations among individuals. If so, then one could argue that this is not a case of group selection after all, because group selection should depend on emergent group features—that is, features that are not simply due to features of individuals. My response would be that this is a matter of terminology. In the flour beetle example, groups make a causal difference. We can decide to reserve the term "group selection" for cases that satisfy the further criterion that emergent group features are involved. Or we can decide to use "group selection" in the broader sense. The first option may imply that the flour beetles exemplify a special case of individual selection that essentially involves groups. The second option would imply that we are dealing with group selection. In either case, we should distinguish this case from ordinary individual selection. As long as we care to make the distinction, the terminology used does not make much difference.

Group Selection (1) is best represented by theoretical work on altruism (Sober and Wilson 1994, 1998; Wilson and Sober 1994). Altruism in the biological, evolutionary sense is behavior of an individual that increases the fitness of other individuals while decreasing the fitness of the individual itself. Altruism cannot evolve through individual selection; however, it is promoted by group selection.

If groups differ in the relative frequency of altruists, then the mean individual fitness of groups with more altruists will be higher than the mean fitness of groups with fewer altruists. Within groups, altruism would be disadvantageous; among groups it would be advantageous. Altruism thus evolves if, on balance, group selection overrides individual selection. Sober and Wilson do not envisage permanent groups. For altruism to evolve, temporary associations of individuals suffice.

Sober and Wilson (1994), however, do not subscribe to the definition of Group Selection (1). Instead, they offer the following definition: "The group was a unit of selection in the evolution of trait T iff one of the factors that influenced T's evolution was that T conferred a benefit on groups" (p. 536). They reject the other definition because it would be overinclusive. They illustrate this with the following example (pp. 544–545). Suppose, in the evolution of traits A and B, the advantage goes to the common trait. Then, if A is more common in one population, and B is more common in another population, the populations will evolve in different directions. But this is frequency-dependent selection, not group selection. For group selection to occur, individuals within a group must be in the same boat. I would argue, in this case as well, that this is a terminological matter. Groups matter causally in this example. As long as we are clear about the processes involved, the choice of terminology, a special case of individual selection involving groups versus group selection, appears to be arbitrary.

Wimsatt (cited in Sober & Wilson 1994) has offered yet another definition: "A *unit of selection* is any entity for which there is heritable *context-independent* variance in fitness among entities at that level which does not appear as heritable context-dependent variance in fitness . . . at any lower level of organization." Sober and Wilson reject this definition as it would lead to the conclusion that their example of altruism would not qualify as group selection.

The various definitions of "group selection" are meant to explicate intuitive notions of group selection. If different authors respond in different ways to a particular example, they must have different intuitive notions in mind. Formal explications cannot then help us to solve controversies, because they thrive on different intuitions. So we have to understand the rationale of the intuitions. If we try to reach understanding by formal explications, we end up running around in circles.

The best way to proceed would be to draw up, as comprehensively as possible, a list of examples of selection processes that essentially involve groups. Perhaps researchers will ultimately manage to agree about a subset from the list of processes that deserves the honorific title of group selection. If not, then we can happily theorize about the different items on the list without inventing overarching categories.

The potential examples of group selection I considered are mostly at centre stage in the literature, but they do not exhaust the possibilities. At least three additional possibilities should come to mind. First, kin selection may count as group selection under some definitions. Kin selection results from a preferential altruistic treatment of kin, which is explained by shared genes (Hamilton 1964). It is commonly regarded as a form of individual selection, but Sober and Wilson regard it as a special case of group selection (Sober and Wilson 1998; Wilson and Sober 1994). Second, I wonder if nonrandom mating in sexual reproduction should count as group selection under some definitions. Fox Keller (1987) has in any case shown that the impact of sexuality is undervalued in population genetics, which for ease of theorizing often focuses on asexual reproduction. Third, groups may affect the outcome of selection processes even if no relevant differences between groups exist. Foraging in groups, for example, may benefit individuals since it fosters predator detection. Thus, we should compare fitnesses of solitary individuals with fitnesses of individuals in groups. I would conjecture that, on many definitions, the distinction between individual selection and group selection breaks down in this case.

Considering species selection, we also face terminological problems. Species selection, from the most common point of view, involves differential speciation or species extinction; this is an analog of Group Selection (2). But not all cases of differential speciation or extinction need to be regarded as species selection. Thus, if the features characterizing species represent an aggregate of features of individual organisms, then organismic selection might account for differential speciation and/or extinction. Vbra (1989) has coined the term "species sorting" for this situation. Her "effect hypothesis" states that organismic selection may cause differences in speciation and/or extinction. She reserves the term "species selection" for processes that do not represent sorting, since they involve "emergent" features of species.

Grantham (1995) notes that the term "emergence" is used in two different senses in the species-selection debate. Vbra uses it for features that cannot be attributed to lower-level entities. Thus, the frequency of brown eyes in a population would be emergent even though the presence of brown eyes is a feature attributed to organisms. Grantham opts for a stricter definition due to Damuth and Heisler (1988), which reserves the term "emergent" for features that cannot be obtained by measurements on individuals. He defends the thesis that emergent features, thus conceived, are not necessary for species selection to occur.

A classic example, which Vbra interprets as "effect macroevolution" (species sorting, not species selection), is her study of South African mammal clades. She argues that, due to higher speciation rates, specialist species in various clades have become more common in the past than have generalist species. She regards this as an example of species sorting, as specialist and generalist feeding habits are aggregate features of individuals. Organismic selection, according to her, suffices to explain the sorting.

Grantham (1995) does not agree: "Although Vbra's explanation does not introduce a higher-level process, I would maintain that this explanation is not reducible [to organismic selection]. The explanation is irreducible because it appeals to differences in a component of species-level fitness (speciation rates) that cannot be reduced to the organismic level" (p. 309). In a footnote, Grantham elaborates on this by noting that many different traits at the organismic level and at the level of populations may affect speciation rates, so that it would be difficult to express the concept of speciation rate in organismic terms.

Some cases of species selection due to different extinction rates are different, according to Grantham. He briefly discusses the example of end-Ordovician extinction among the trilobites. This extinction was allegedly caused by a drop in ocean temperature which negatively affected species with pelagic adult lifestyles and species with planktotrophic larvae. This is a case of effect macroevolution. Differences in particular organismic traits led to differential extinction. But the extinction of species simply amounted to the death of all individuals belonging to the species, death being caused by traits of the individuals which did not fit in with the environment.

What Grantham states in the footnote is misleading. The fact that "many . . . traits . . . may affect speciation rates" does not imply that "it would be difficult to express the concept of speciation rate in organismic terms." It is in fact easy to define the concept of speciation (and the concept of speciation rate) in organismic terms. A speciation event has occurred if descendants of a single species belong to two different species. Thus "speciation" is defined in organismic terms if "species" is so defined. But that should not be difficult. The biological species concept, for example, views species as collections of interbreeding organisms reproductively isolated from other collections. This is an explication in terms of individuals and relations among them.

Considering the causes of "fitness" at the species level that affect speciation and extinction rates, I agree with Grantham that these come in many different kinds, at least in the case of speciation rates. But that is not the issue. If species selection is the analog of organismic selection, then causes of fitness need not enter the picture on a common modeling of selection. Models of population genetics for organismic selection likewise disregard causes of fitness differences. Full-fledged explanations of evolution involving (organismic) natural selection should appeal to these causes. But here, at the organismic level, causes also come in many different kinds, so that we cannot generalize over them. Thus full-fledged evolutionary explanations need to resort to natural history in some respects. This carries over to the explanation of events at the species level. Now this does imply that we cannot get at "reductive" explanations that identify *general* causes of speciation at the organismic level. But that in itself does not warrant a special status for species selection. We have no explanations that identify general causes at the organismic level either. The fact that we cannot provide general causal explanations for any level of organization is a negative that cannot be taken to imply that processes are taking place at several levels.

If the concepts of species and speciation (and extinction) can be defined through the organismic level, as I argued, then we are entitled to describe speciation as a process involving individuals. Would this imply that species selection is ultimately composed of organismic processes? Indeed, it may imply this. But, at the same time, the nature of the organismic processes depends on how the organisms are organized into species. The fact that organisms are organized into higher-level "wholes" called species makes a huge difference for evolution.

For the rest, we should take heed here not to confuse the thesis that species selection is composed (in all cases or in some cases) of organismic processes with the stronger thesis that it is composed (in these cases) of organismic *selection* processes. The latter thesis is false. It is conceivable, for example, that allopatric speciation takes place due to genetic drift in geographically separated populations. This shows that a component of "selection" at the species level may result from drift, not selection, at the organismic level.

Selection at the species level and organismic selection are different also for other reasons. Grantham, like most authors writing about species selection, departs from the analogy that construes both organismic selection and species selection as differential survival of entities. So far, I have taken for granted that the analogy makes sense. In fact, it is misleading. First, in organismic selection models, reproduction does not lead to new types of entity. Relative frequencies change, but the types remain the same. But by definition the "reproduction" of species through speciation generates entirely new types. Thus it would be appropriate in some respects to analogize speciation with mutation rather than selection or a component of selection (Sober 1984: 360–361). Speciation, like mutation, is a source of new variants. Second, the selective disappearance of types due to selection has very different ramifications in the two situations. In organismic selection, a disadvantageous trait may disappear from a population due to differential mortality of phenotypes while most other traits persist. In species selection, extinction implies that all traits of a species, not merely some disadvantageous ones, are lost.

Seemingly convincing examples showing that species selection is a separate process have been offered by Stanley (1979). Sober (1984: 366–368) elaborates one of the examples as follows. Suppose that we begin with two species of grasshopper: a winged species and a wingless species. Suppose further that the winged species, by virtue of being winged, establishes few isolated populations, so that the speciation rate is low, while the wingless species produces many daughter species. It is conceivable that the presence or absence of wings does not affect the fitness of individuals, and that total numbers of winged individuals and of wingless individuals stay the same. Thus, we get many small wingless species and few large winged species. That would amount to species selection in the absence of organismic selection.

Would this represent selection for winglessness? Yes and no. At the level of organisms, no selection for this property occurred. But if species instead of organisms are our "benchmark," selection for winglessness did occur. Sober approvingly refers to Eldredge in this context. Eldredge has remarked that the emphasis in evolutionary theory has been overmuch on the transformation of characters within a lineage at the expense of the origin of new taxa in cladogenesis.

All this is compatible with the view that macroevolution does result from processes at the organismic level. We can describe the Stanley–Sober example in terms of these processes. Speciation in the example involves geographic isolation. But geographic isolation as such is not sufficient. Speciation occurs when the isolated populations diverge through drift or different selection regimes. Let us assume that different selection regimes are causally responsible in the grasshopper case. Then organismic selection would cause the divergence ultimately leading to speciation. We should not forget here that speciation primarily concerns divergence in features not mentioned in the example as Sober describes it. The presence or absence of wings need not play a role in any process of divergence. *This* feature is responsible merely for another aspect of the speciation process—namely, geographic isolation.

We saw that Grantham (1995) holds that emergence in a strict sense is not necessary for species selection to occur. He appears to hold that emergence in a broader sense is necessary, though. Stidd and Wade (1995) have argued that emergence is not necessary at all. They consider a situation in which some species are composed entirely of black organisms, while other species are composed of white organisms. If black organisms were to be at a disadvantage, we would get differential species extinction due to this. Vbra regards this as species sorting, not species selection. But Stidd and Wade argue that it is a case of species selection. Selection at the level of individuals is defined as a process that takes place within species. In this example, we are dealing not with intraspecific selection, but with interspecific selection.

Again, I can but conclude that this is a terminological matter. The fact that we are dealing with different species makes a causal difference. The extinction of black morphs within a population is different from the extinction of black morphs that constitute an entire species. Nonetheless, the extinction of the species is explained by selective deaths of individuals. We may choose to call this differential species extinction due to individual selection or to species selection. Nothing much hinges on this.

The upshot of this section is that groups and species can influence selection processes in different ways. The terminology (group selection, species selection) with which we describe these influences is not important as long as we take care to distinguish between them. Because the influences come in different kinds, highly general claims—for example, about the relative importance of group selection and individual selection—are futile. We have to resort to lower levels of generality—that is, to natural history.

5.5 CONCLUSIONS

Natural selection leads to changes in genes, genotypes, and phenotypes. It has implications for individuals, groups, and species. Yet this need not imply that it "acts on" all these "units." Which are the units of selection? A large, confusing literature has aimed to answer this question.

In fact, we are dealing with two different issues. In "ordinary" selection, genes are differentially replicated due to organisms interacting with their environment. Thus we should distinguish between two selection processes: replication and interaction. One issue is whether the focus in theorizing should be on replicators, or on interactors or "vehicles."

My analysis indicates that this should be a nonissue. A comprehensive representation of any selection process should account for both replicators and interactors. It is true that mathematical equations of population genetics may refer to genes only, but the equations as such do not represent selection processes. A full-fledged interpretation of the meaning of the equations unavoidably refers to interactors also.

Casual inspection suggests that the equations of population genetics are laws of nature. But the equations as such have no empirical content. If we make the context of interpretation explicit, we get natural history rather than laws.

The second issue concerns the nature of the interactors. Is it possible for selection to occur not only at the level of organisms, but also at the level of groups or species? I have argued that the existence of groups within species, and the existence of species, does causally affect the evolutionary process. Groups and species may be involved in the causation of evolution in many different ways. The overall picture has the flavor of natural history, not highly general theory.

In what situations should we use the labels of group selection and species selection? Different authors answer this question in different ways. In my view, the entire controversy is, to a large extent, a matter of linguistic convention. We should be clear about the processes we are considering. Overarching labels attached to the processes are mostly arbitrary.

6

Evolution and Altruism

6.1 INTRODUCTION

Is human behavior exclusively motivated by self-interest? Common sense indicates that we should deny this. Some behaviors are so motivated, other behaviors are not. It all depends on the persons concerned and the situations they find themselves in. Yet the doctrine of universal self-interest—psychological egoism, for short—has gained the support of many researchers in science. Researchers in several disciplines hold that the main support comes from evolutionary biology. I criticize this view in the present chapter.

First, I argue that the thesis of psychological egoism does not make sense without specification, because terms such as "altruism," "egoism," and "selfishness" represent different concepts. The thesis is so unclear that it would be unwise to take sides in disputes over it. The same is true of the thesis of ethical egoism, which applauds enlightened self-interest as a normative guideline. Each thesis actually represents a heterogeneous collection of views that belong to dissimilar contexts. If we attend to contextual detail, then we cannot fail to notice that human social behavior comprises genuine forms of altruism, egoism, and much else that does not fit either label.

I use conceptual analysis to show that "altruism" and "egoism" are multiply ambiguous labels. Next, I demonstrate that ambiguities explain why ethicists have not managed to reach consensus over ethical egoism. Furthermore, I argue that theorizing about the issue is also problematic because of lack of contact with empirical science.

Many researchers have argued that empirical science supports psychological egoism. Their emphasis is often overmuch on simplified versions of evolutionary biology that allegedly imply that behavior in animals and man is pervasively selfish. Biology, however, does not support this view of selfishness.

I review biological research that helps us dispense with simplified versions of evolutionary biology. Next, I criticize evolutionary psychology, which draws on these versions to defend psychological egoism. Finally, I comment on the views of Neven Sesardic (1995), who has provided the most extensive survey of possible strategies to reconcile altruism with evolutionary theory. He argues that few strategies are viable. But Sesardic's philosophical analyses should be rejected, since they depend on overly general concepts. We should not blame Sesardic in particular for this. His conceptualization of the problem fits well with extant philosophical approaches of altruism. Misguided generality is indeed a pervasive feature of philosophy, not least ethics (Burian, Richardson, and van der Steen 1996; Kirkham 1992; van der Steen 1993, 1995; van der Steen and Musschenga 1992).

As you will have noticed, I use the terms "egoism" and "altruism" in two different senses. They refer to certain behaviors and also to theories dealing with these behaviors. Intended meanings should be clear from the context.

6.2 EGOISM AND ALTRUISM: A SCHEME FOR CONCEPTUAL ANALYSIS

The classification of behaviors or acts into altruistic and egoistic ones is difficult, as many different criteria of classification are potentially involved. In ethical writings on egoism and altruism, the following criteria for classifying behaviors or acts commonly play a role:

1. Benefits for self versus no benefits for self.
2. Benefits for others versus no benefits for others.
3. Motivated by benefits involved versus not so motivated.

I have kept the list simple. Notice, for example, that Criterion 3 is in fact a composite one. In a particular case, there may be benefits for self and for others, whereas motivation may concern benefits for self only, or benefits for others only. Even this simplistic scheme would lead to a classification with at least eight fully specified categories of behaviors or acts. This should dispose of any simple dichotomy of altruism versus egoism.

The notion of benefit calls for further distinctions. First, I would distinguish between *first-order* benefits and *second-order* benefits. Second-order benefits involve attitudes toward first-order benefits. Getting money is a first-order benefit. Feeling good about getting money would be a second-order benefit. In preventing another person from drowning, and feeling good about it, you realize a first-order benefit for the other person and a second-order benefit for yourself.

Second, there are *primary* benefits and *secondary* benefits. If I go around helping people, the result may be that I am generally liked. That would amount to a primary first-order benefit for others and a secondary first-order benefit for myself

The two distinctions just introduced have a formal character. We also need material distinctions that concern the *currency*. For example, evolutionary biologists concerned with altruism and egoism concentrate on reproductive output (numbers of descendants) as an overriding criterion for benefits. This is understandable in view of the context in which they are working. Ethicists seldom use this criterion. They work with different currencies—for example, satisfaction, pleasure, happiness, and various forms of utility.

In addition to all this, we should recognize that the general category of no benefits is simplistic. Some acts without benefits for anyone may be neutral and harmless. Other acts cause harm. Human behavior throughout history has been rife with acts of senseless violence that did not serve any interest at all. Likewise for acts out of envy, spite, jealousy. This shows once again that ethical theories pitting egoism against altruism foster a biased view of human nature.

The general category of motivation is likewise heterogeneous. Motives can have different contents. Apart from this, the concept of motive can range over a continuum from reflective deliberation to subconscious motivation. Furthermore, an overarching concept of motivation easily evokes the misleading picture of motives as inner forces that causally propel acts. The picture makes sense in some situations, but as an overall frame of reference it is problematic. A cry for help from a child in the river may cause you to jump in the water and rescue him or her. Do we need to postulate an inner motive as an additional force that guides your behavior? I would not understand what motive-talk refers to in this case.

The concept of self in discussions of egoism and altruism is not perspicuous either. In empathic exchanges among persons, the sense of self dissolves. Self-interest is at odds with such exchanges because the concept of self-interest ceases to be applicable.

In discussions within ethics, the criteria of classification, and the further distinctions I introduced, all play a role. However, in most publications only a limited subset of them is considered. Different publications are often concerned with different subsets. This easily leads to confusion and spurious disagreement.

"Altruism" and "egoism," in brief, are placeholder concepts that, for theorizing to make sense, need to be replaced by more specific notions. Theorizing will have to take the form of natural history, and it is futile to search for general laws about altruism and egoism.

Most authors would agree that the core meaning of "egoism" involves benefits for self and motivation directed toward such benefits. Altruism, contrariwise, involves benefits for others and motivation directed toward such benefits. Beyond core meanings, many different characterizations are possible. In the next section, I show by examples from ethical writings that this fosters bias and spurious disagreement.

6.3 EGOISM VERSUS ALTRUISM: SAMPLES FROM ETHICS

Considering egoism and altruism, ethicists often indulge in abstract theorizing that disregards science. The philosopher Thomas Nagel (1970) has indeed argued that this is how it should be. In a famous book, *The Possibility of Altruism*, he offers a classic defense of altruism in ethics, in which science is not assigned any role. The following passage indicates how he defends this view.

> The account I offer will depend on a formal feature of practical reasoning which has a metaphysical explanation.
>
> Alternative hypotheses fail as plausible candidates for a complete account of altruistic action because none of them provides the type of simple, absolute generality which is required. There is a considerateness for others which is beyond the reach of complicated reflections about social advantage, and which does not require the operation of any specific sentiment. The task is to discover an account of this general, passionless motivation which will make its existence plausible. Introspective and empirical investigation are not very useful in this area since the motivation is often partly or completely blocked in its operation by the interference of corrupting factors: repression, rationalization, blindness, weakness. Arguments and theoretical considerations can, however, reveal the form of an altruistic *component* in practical reason, which will be one contribution among others to the genesis of action. [Nagel 1970: 82]

The first thing to notice about Nagel's argument is his emphasis on simplicity and generality. Accounts of altruism fail, according to him, if they are not simple and absolutely general. I would argue that simplicity and generality are by no means the only methodological criteria bearing on theories. Our theoretical terms should first and foremost satisfy the criterion of clarity. Considering the term "altruism," we face a dilemma. The term is intolerably unclear precisely because it is simple and general. If we opt for clarity, we had better sacrifice generality and replace the overarching label of altruism by an array of more specific labels.

How can Nagel justify his rejection of data as evidence? The sole reason he mentions is that corrupting factors may interfere. If that were a good reason, then we had better stop doing empirical science altogether and turn to armchair philosophy, because "corrupting factors" potentially play a role in all empirical research.

Common sense and science combine to show that human beings exhibit particular varieties of altruism in particular situations. If we want to know more about this, we should attend to the particulars. Nagel's claim that altruism-in-general is possible for theoretical reasons does not add to what we know about particulars.

Many ethicists would reject Nagel's view on the ground that ethical egoism fits in with morality if egoism is given a broad interpretation. The idea is that we may serve our interests by apparently selfless acts. Thus, self-interest could provide us with reasons to be moral. I briefly consider two representative examples of theorizing along these lines.

The following passage is from an article by David Gauthier (1987):

> Suppose that whether or not one accepts social requirements as overriding reasons [i.e. reasons to be moral] affects the regard one receives from others—in particular, their willingness to accept one as a partner or participant in mutually advantageous interactions. If one accepts social requirements as overriding one's self-anchored reasons, then one may be relied upon to do one's share in activities and practices that work for the good of everyone alike, even if one could benefit oneself even more by free-riding on the efforts of others. Whereas if one does not accept social requirements as overriding, then one may be expected to free-ride whenever this seems individually most beneficial. Those who may be relied upon not to free-ride are welcome partners; free-riders are not. . . .
>
> . . . We must distinguish clearly between persons who act only on self-anchored reasons, and so adhere to social requirements only when it is beneficial for themselves to do so, and those who, for self-anchored reasons, dispose themselves to adhere to social requirements whether or not it is actually beneficial for them to do so. Which kind of person does one have better self-anchored reason to choose to be? Our present argument leads to the conclusion that one has better reason to choose to be the latter kind of person, who is disposed not to free-ride even though free-riding would of course benefit her. [Gauthier 1987: 19–20]

At first sight, Gauthier is arguing here against full-fledged ethical egoism, because he assumes that self-anchored reasons cannot suffice as a basis for morality. In some circumstances, social requirements override such reasons. However, social requirements, in Gauthier's example, override self-anchored reasons *for self-anchored reasons*. The implicit point of Gauthier's argument is that two kinds of benefit play a role in the example—primary ones and secondary ones. Becoming a welcome partner provides a motive for letting social requirements override self-anchored reasons and allied primary benefits, but becoming such a partner is also a benefit, a secondary one. That is why I would regard Gauthier's view as a *defense* of ethical egoism.

Gauthier's line of reasoning suggests that secondary benefits for self associated with primary benefits for others could justify a broadly defined form of ethical egoism.

Scott (1988) approaches the issue in a different way:

> . . . I take up a distinction, for which I have coined terms, between *direct* and *reflexive* purposes. An action is directly self-interested if *its purpose* is a good (as a means or as an end) for the agent. It is reflexively self-interested if *his doing it* is a good for him, whether its purpose is or not. . . .
>
> The point to which the distinction moves is that a selfless life is one whose direct aim is selfless. If it is also reasonable, its reflexive aim is self-interested. I am not saying that the agent's motive must entirely consist in the reflexive good, but I am saying that a selfless direct motive makes no sense by itself. [Scott 1988: 493–494]

Scott also appears to defend egoism in a broad sense of the term, but he focuses on second-order benefits. According to him, arguments like the one in the passage quoted justify two theses: (1) "The only reasonable, conclusive motive for choosing a certain kind of life is that it is a good for oneself," and (2) "A life of human decency is always a good for the person who leads it" (Scott 1988: 482).

Is it possible to justify ethical egoism by giving it a broad interpretation? Arguments to this effect, like the ones presented by Gauthier and Scott, have a plausible ring. However, broad interpretations run the risk of becoming trivial because self-interest is easily made a defining feature of human acts.

If we take a broad view of benefits to include secondary and second-order ones, every act appears *necessarily* egoistic in the sense that people can but do what they want to do, so that their acts will always be self-serving in some way. Many authors have noted that this would make psychological egoism a boring view since it would become true by definition, and that it would make ethical egoism pointless. Taylor (1975) formulates the issue as follows:

There is one argument given by the psychological egoist, however, which rests on a semantic confusion and not on factual claims about human motivation. Suppose an egoist is presented with a [paradigm case of "altruism"]. The egoist might then say the following about such a case: "Granted that in doing this act the person did not have what most people would call a selfish motive, or even a motive of self-interest. Still the act was a voluntary one, and all voluntary acts have some motive behind them. By doing the act the person was satisfying the motive. Hence he did gain satisfaction from doing the act. He would not have done it if no satisfaction of any of his motives would result from his doing it. Consequently the act served his self-interest after all, since it was done to satisfy whatever motive he did have in doing it."

Here the egoist has made a basic change in his position. For he is now claiming that the satisfaction of *any* motive is to be taken as self-interest, whereas what he had been saying before was that all action is motivated by a certain *kind* of motive, the kind we would all classify under the general category of selfishness or self-interest. [Taylor 1975: 41]

Taylor's point is that an overinclusive conception of benefits or interests (and of motives) turns egoism into a trivial, tautologous doctrine. This indicates that doctrines of ethical egoism will be meaningful only if some *currency* in which to measure benefits is specified. Doctrines of psychological egoism that use the term "egoism" in highly general ways are indeed meaningless.

The well-known doctrine that it is rational to be moral comes close to being a tautologous version of psychological egoism. Rescher (1987) provides a representative defense of this doctrine. He states that rationality is a matter of seeking optimal resolutions to the problems that we face, pursuing appropriate objectives, acting for the best. This is so "by definition," as it were, says Rescher. Self-interest is a sensible objective for people; so it is a guiding factor for rationality. It seems to me that this, too, is virtually a matter of definition if self-interest is broadly defined. Next, Rescher states that people generally make interests of others part of their own. Now, heeding the interests of others is the crux of morality, says Rescher. I take it that this is again a matter of definition. Hence, we can infer that it is rational to be moral.

Rescher summarizes his main line of reasoning as follows.

Premise 1: The intelligent cultivation of one's real self-interest is quintessentially rational.

Premise 2: It is to one's real self-interest to act morally—even if doing so goes against one's immediate selfish desires.

Conclusion: It is rational to be moral.

I would regard the first premise as a definition. The second premise is an unspecific empirical assertion. The conclusion is a restatement of the second premise contingent on the definition in the first premise. All in all, Rescher's thesis that it is rational to be moral is practically empty because it is based on an unspecific empirical claim and unspecific definitions of key concepts such as rationality and morality. The definitions take off from an undefined, unhelpfully broad notion of interest.

The case against general views of ethical egoism is also supported by the curious fact that they allegedly include contradictions in addition to tautologies. This suggests that something is seriously wrong with general views.

Critical comments on "high-minded egoism," as formulated by Lemos (1984), illustrate the contradiction problem. High-minded egoists would aim to act virtuously and develop a good character because they regard these things as benefits for themselves. Lemos suggests that for *conceptual* reasons, high-minded egoism is impossible. The following passage explains this:

If a man were motivated solely by the desire that he act friendly, but not by a desire for the well-being of his friends, then such a person would not be friendly and we should not say that his act was motivated by friendship. If a man is motivated solely by the desire that he be virtuous, but not by any of those other characteristics of his act which would make it virtuous, then we would not properly call his act virtuous. On this view, the goodness or moral worth of an act of friendship depends, for example, on its being motivated by a desire for the well-being of one's friends. Similarly, the goodness or moral worth of an act of virtue depends on its being motivated by certain features of the act apart from its being an instance of one's acting virtuously. [Lemos 1984: 554]

In a subsequent analysis, Lemos shows that a strong form of ethical egoism, which aims to put *purely* selfish motivation at the basis of moral rightness, is logically impossible. This is remarkable because, from Taylor's analysis, egoism as a general doctrine appears to be tautologous under a different description.

Let us take stock. Concerning ethical egoism as a general doctrine, four different stances are possible in principle. The doctrine could be (1) valid for substantive reasons, or (2) invalid for substantive reasons, or (3) valid for logical reasons, or (4) invalid for logical reasons.

These options are continually discussed in the literature. This suggests that the issue is approached in the wrong way. The scheme presented in section 6.2, which covers major distinctions in abstract discussions, shows that "egoism" and "altruism" as general terms cover many different concepts. Thus, differences of opinion over the options that are logically possible will often reflect different conceptualizations.

In research on egoism and altruism, our priority should be to get rid of options (3) and (4). Ethics should be concerned first and foremost with substantive issues. The fact that tautologies and contradictions have long since been subjects of research concerning egoism and altruism suggests that we are dealing with unprofitable analyses. In the next section, I outline a more profitable approach that moves beyond such analyses.

6.4 THE RELEVANCE OF EMPIRICAL ISSUES

The scheme introduced in section 6.2 indicates that general concepts of egoism and altruism are useless. The tenaciousness with which ethicists keep using such concepts calls for an explanation. An important cause of the tenaciousness appears to be that ethics undervalues empirical matters.

General concepts of egoism and altruism cover heterogeneous categories. The heterogeneity is at odds with the requirement that concepts should have an unambiguous empirical reference. Normative views of egoism and altruism presuppose that the concepts of altruism and egoism apply to behaviors in the real world. To be relevant empirically, the concepts should exhibit homogeneity of reference to a reasonable degree.

General doctrines of psychological and of ethical egoism are problematic because their conceptual anchors to the real world are not sufficiently natural and specific. Adam Smith and David Hume recognized this long ago. Their analyses are relevant empirically since they discussed normative matters in the context of a rich moral psychology. A thought-provoking historical essay by Holmes (1990) shows that their analyses of self-interest are much more sophisticated than abstract contemporary work.

Hume himself was echoing Shaftesbury, Hutcheson, and others when he denounced the false "love of simplicity" underlying motivational reductionism. To say that patriots and misers, cowards and heroes all aim exclusively at "their own happiness and welfare" illustrates only how little we can learn about behavior by adducing self-interest. We can always *say* that the altruist includes the welfare of others in his own utility function. But when the motivational reductionist traces all action to self-love or the rational pursuit of personal advantage, he "makes use of a different language from the rest of his countrymen, and calls things not by their proper names." Theorists are free to indulge in such linguistic idiosyncrasy, of course. If they make no distinctions, then, naturally enough, everything will be the same. [Holmes 1990: 269]

Smith and Hume recognized that the use of overgeneral concepts may generate tautologies. They opposed simple dichotomies of egoism versus altruism. Destructive behaviors generated by irrational passions are not covered by such dichotomies. Against the background of such behaviors, self-interest may be seen as a positive force. Indeed, Holmes argues that "the postulate of *universal* self-interest, although logically incompatible with insight into the rich variety of human motives, first rose to cultural prominence because of its unmistakably egalitarian and democratic implications" (Holmes 1990: 268). The emphasis on self-interest in the present Western culture has lost this positive flavor.

Empirically relevant *concepts* are necessary for ethics to be anchored to the real world. If ethics is to provide codes of conduct for humans, it also needs empirical *theses* indicating what human beings are like. This should be clear even for theorizing at the general level that I have been dismissing. If universal self-interest were part and parcel of human nature, we would not need to encourage or justify it. Thus, ethical egoism does not sit well with the assumption that psychological egoism is generally valid.

Psychological egoism is indeed taken for granted nowadays in many areas of empirical science. For example, Wallach and Wallach (1983) have argued convincingly that psychology and psychotherapies commonly presuppose—wrongly so—that human beings are pervasively selfish. How odd that ethical egoism as a general doctrine gets much attention at the same time.

Once we reject overgeneral doctrines concerning egoism and altruism, we can feel free to embrace empirical views and normative views concerning specific forms of altruism geared to specific situations.

Many persons would help a victim of an assault or an accident. Should the helping behavior be regarded as a form of altruism? That clearly depends on the underlying motivation. If the motivation were selfish, then the label of altruism would not be appropriate. Two competing models could in principle explain the helping behavior. According to the negative-state relief model, helping a victim can alleviate our own negative emotions. This may provide the actual impulse to provide help. The empathy–altruism model postulates empathic concern as the basic motivation. Experiments indicate that empathic concern is indeed a common motivation in helping behavior (Schroeder, Dovidio, Sickiby, Matthews, and Allen 1988). This indicates that the thesis of psychological egoism is not generally valid.

Next, consider marriage. Concerning marriage and other close relationships, Wallach and Wallach (1983: 178) note that "conditionality as the key to keeping one's relationships healthy and happy, is a theme sounded by many researchers." In the terminology of section 6.2, conditionality involves secondary benefits for self that motivate us to aim at primary benefits for others. Wallach and Wallach show that researchers in psychology commonly regard this form of selfishness as rational and commendable. It is promoted in many psychotherapies. Therapists who take this line foster a form of ethical egoism that they suppose works out well in this context.

The implication is that, to make the relationship work better, each partner has to be assured of deriving sufficiently favorable net benefits from maintaining it. Neglected is the possibility that instead the very terms of the analysis—as applied both by the couple and by the psychologist—are at fault. What may be wanting is not more attention to balances of returns but an approach to relationships that views another's welfare and commitment to furthering it as primary considerations. [Wallach and Wallach 1983: 179]

Wallach and Wallach rightly deplore this attitude. Evidence shows that conditionality is pronounced in the reciprocation of negative affect in distressed couples. Happy marriages instead thrive on commitment and trust. A specific form of ethical egoism is proved wrong in this case by empirical data.

At first sight this seems odd. How could we reject a normative thesis on empirical grounds? The answer is simple. Normative theses commonly presuppose the truth of empirical ones. Psychotherapists who promote ethical egoism presuppose that it works out well. It does not—not as a rule, anyway.

Wallach and Wallach review many other examples. Their analyses support the following conclusions. Genuine forms of altruism are common in social interactions. Yet, overall, psychological theory in our century has been guided by a biased emphasis on egoism. Psychoanalysis, humanistic psychology, social psychology, and developmental psychology are all biased in this way. The possibility that human beings often engage in genuine forms of altruism is seldom investigated in a methodologically proper way. The pervasiveness of self-interest is presupposed rather than demonstrated.

6.5 EXPLAINING ALTRUISM IN ANIMALS

Ethical egoism—the thesis that enlightened self-interest is morally commendable—and psychological egoism—the thesis that human behavior is indeed motivated by self-interest—are often reinforced by evolutionary considerations. I argue that these considerations are as suspect as the theses in themselves. First, I review biological research about egoism and altruism.

In evolutionary biology, the terms "altruism" and "egoism" have meanings related to fitness—that is, reproductive interests. A behavior is altruistic if it increases the fitness of the recipient and decreases the fitness of the organism exhibiting the behavior. The terms "egoism" and "selfishness" represent the converse relationship.

Altruism thus defined is seemingly at odds with evolutionary theory. Natural selection should promote egoism and inhibit altruism. Yet, on the face of it, many organisms exhibit altruistic behaviors. Biologists have developed three theories to account for the discrepancy.

Evolution and Altruism

First, it is conceivable that altruism is generated by group selection overriding individual selection. Group selection is not very popular nowadays, but the tide seems to be turning (Sober and Wilson 1994; Wilson and Sober 1994).

In addition, from a genic perspective, altruistic behavior toward relatives may be favored by selection since it promotes the spread of genes identical to genes of the organism exhibiting the behavior. Thus, we must replace the ordinary fitness concept by a concept of inclusive fitness which takes this into account. The ensuing selection process is called kin selection (Hamilton 1964).

Finally, reciprocal altruism is evolutionarily feasible (Trivers 1971). Altruistic behavior is beneficial for the organism exhibiting it if recipients reciprocate. Reciprocal altruism works well only if reciprocation does occur. Hence it calls for protection against cheaters.

On the kin-selection account and the reciprocal-altruism account, altruism involves benefits in the form of enhanced fitness for the altruist. Thus, altruism reduces to egoism or to mutualism which benefits all concerned. Barring the exception of group selection, biology appears to saddle us with a paradigm of universal self-interest.

The implied emphasis on self-interest thrives on a misleading conceptualization. Altruism and egoism, in narrow biological senses, should represent endpoints of a continuum, with mutualism in between. Putting mutualism into the category of egoism generates stakes in favor of egoism since the term "egoism" gets a broad definition, the term "altruism" a restricted definition. When biology is applied to domains where researchers may not be familiar with technicalities of biological notions, the terminology may cause bias.

The three accounts of altruism considered are prominent in some biology texts that focus on evolution and behavior. Researchers in other disciplines drawing on biology tend to regard them as an exhaustive set. That is misleading as biologists have uncovered behaviors that call for different explanations.

Participation in charity is a paradigmatic example of altruism in human beings. But *Homo sapiens* appears not to be the only species that exercises charity. The Arabian babbler *Turdoides squamiceps*, a group-breeding songbird, cherishes charity even more than do human beings. The babblers even fight over opportunities to participate in charity. For example, the birds help nonrelatives to feed nestlings, they feed other adults, and they help other babblers defend their territories, though dominant babblers prevent subordinates from helping the group. Zahavi (1995) argues that kin selection or reciprocal altruism fail to explain this odd phenomenon; presumably, he would also rule out group selection. He proposes that the birds that manage to help other babblers benefit by way of social prestige. They invest in wasteful characters to advertise superior qualities. This attracts collaborators and deters rivals. Helping thus enhances fitness.

Apparent altruism is again reduced to self-interest, in the biological sense linked with fitness. But the example indicates that apparent altruism is more common than theories of biology regarded as paradigmatic by nonbiologists suggest. The behavior of the babblers may be regarded as egoistic in a technical biological sense involving fitness. But, if we use "egoism" to designate notions of psychology, ethics, or everyday life, this does not allow us to infer that the babblers are egoistic.

Foraging in the colonial cliff swallow *Hirundo pyrrhonota* also does not fit theories regarded as paradigmatic outside biology (Connor 1995a). The swallows forage on insects in the air. If they come across a swarm of insects, they emit calls alerting conspecifics that food has been located. From a superficial interpretation, the calls represent altruism, because the birds would do better if they kept the insects for themselves. However, they actually benefit from their behavior. It is difficult to follow insect swarms. Hence, foraging in groups benefits all the swallows. The calling birds are not behaving altruistically. Neither are their conspecifics behaving selfishly. The conspecifics benefit from the calls, but they also help the callers to have a good meal.

This unmasks apparent altruism as a different kind of behavior. While we could choose to regard the behavior as selfish, it is not egoistic in a strict sense. The foraging of the swallows is special in that it benefits all of them, while it does not call for precautions against cheaters. It deserves to be called genuine mutualism. The fact that precautions are unnecessary is noteworthy, because the paradigm of reciprocal altruism suggests that helping nonkin arises only if special safeguards exist against cheating.

Adoption in birds also cannot be explained by group selection, kin selection, or reciprocal altruism. In over 150 species of bird, adults take care of young of conspecifics that are not genetically related to them (Riedman 1982). T. D. Williams (1994), who studied the phenomenon in the lesser snow goose, *Chen caerulescens caerulescens*, argues that it benefits all the geese. Increased brood size confers a higher dominance rank on the parents, which enhances opportunities for feeding. It increases overall vigilance and helps settle conflicts with geese outside the family. The fitness of parents and young including adoptees is thereby increased.

Other examples of mutualism are abundant (Connor 1995b; Dugatkin 1997, 1999; Dugatkin and Mesterton-Gibbons 1996). They show that mutual benefits in behavioral exchanges can come about in many different ways.

The accounts of apparent altruism share the assumption that the behavior of an organism should enhance its fitness. This assumption is problematic. Some behaviors may be selectively neutral. Pleiotropy in which disadvantageous traits are genetically coupled to advantageous traits may foster altruism. Population geneticists would indeed regard as overly simplistic the general thesis that selection leads to the maximization of fitness (Hartl and Clark 1989). Unfortunately, simplistic conceptualizations keep pervading theoretical studies of altruism outside biology.

All in all, biology does not support the thesis that social behavior in animals is pervasively selfish. Behavior comes in different kinds, ranging from egoism via mutualism to altruism. Evolution at most constrains behavior in that altruism

6.6 THE PARADIGM OF EVOLUTIONARY PSYCHOLOGY

Leda Cosmides and John Tooby (1992) have studied cooperation among humans from an evolutionary perspective. They are leading researchers in a discipline called evolutionary psychology. They search for psychological mechanisms explaining human behavior. Variability of behavior notwithstanding, human beings must share underlying mechanisms that are adaptive and genetically anchored. That, according to Cosmides and Tooby, is what evolutionary biology leads us to expect.

The point of departure of Cosmides and Tooby is game theory applied to reciprocal altruism. Game theory is concerned with costs and benefits. In a behavioral exchange between two persons, each person incurs costs and receives benefits. Natural selection promotes behavior that minimizes the ratio of costs and benefits. This suggests that a single, noniterated behavioral exchange between two individuals rarely takes the form of cooperation.

The well-known prisoner's dilemma illustrates this. Suppose two prisoners awaiting trial are in the following situation. Each of them is offered freedom if he or she confesses, provided that the other prisoner does not confess. If the other prisoner does not confess, he or she will get a heavy sentence. If both confess, their sentences will be moderate. If both do not confess, their sentences will be light. The best option would be for both prisoners not to confess. However, a prisoner runs the risk of getting a heavy sentence if the other prisoner defects. Both prisoners are likely therefore to confess. That ensures a moderate sentence. Cheating is probable, and the best outcome to the dilemma is unlikely to be reached.

Cosmides and Tooby (1992: 175) briefly deal with the objection that real life is not like a prisoner's dilemma, because real-life exchanges are simultaneous, face-to-face interactions in which participants can recognize each other's intentions. They regard this as an exception. Simultaneous mutual aid is rare in nature since items of exchange are often part of acts that cannot be undone, and needs and abilities of organisms are seldom complementary. Simultaneous exchange in nature is also often senseless.

A one-move prisoner's dilemma therefore does not allow of cooperation. But in a series of games, cooperation becomes feasible since participants can take into account what a partner did on previous occasions. It is wise to cooperate with cooperators and to exclude cheaters. Cosmides and Tooby side with Axelrod (1984), who has charted the decision rules needed to achieve this. For example, the cognitive program underlying behavior must contain algorithms that are sensitive to cues indicating that an exchange is offered, or that reciprocation is expected. We need many rules to estimate and to evaluate costs and benefits, and a rule causing appropriate punishment upon cheating. Axelrod's list contains twelve rules of this kind. Cosmides and Tooby argue that this is just a modest beginning. Further rules are needed to account for specific situations. For example, the rules by which friends play games will be different from those of strangers. The rules may be operative without being consciously entertained.

Cosmides and Tooby argue that this "computational" theory of cooperation only explains behavior in a specific, social domain. We are not dealing with a general cognitive mechanism of formal reasoning applied to social interaction, but with specific mechanisms. They argue their case with reference to research about the so-called Wason selection task. The following example explains Wason selection.

Imagine you are given the task of checking whether the following thesis is false: "If a person is clever, then they pass their exam." To check this, you get four cards with information printed on both sides. One side says whether a particular person is clever, the other side informs you about his or her exam. The cards are on the table, so that you see one side only. The text you see on the four cards reads: (1) John is not clever, (2) Mary is clever, (3) Peter passes his exam, (4) Irvin does not pass his exam. Which cards should you turn in an effort to obtain evidence that the thesis is false?

You need logic to answer this question. The thesis is a statement with the form "If p, then q." Now, p can be true or false, and q can be true or false. That yields four possibilities, only one of which implies that the statement is false: p is true and q is false. Therefore, only cards (2) and (4) could show that the thesis is false.

If you did not find the right solution, do not conclude you are stupid. Surprisingly many people fail in this task even if they pass their exams with honors. Human reasoning apparently need not comply with the canons of logic.

Tests involving Wason selection tasks suggest that general-purpose cognitive mechanisms are rare, if they exist at all. Instead, we find domain-specific mechanisms with a particular content in addition to formal features. The content of statements to be evaluated in tests affects the results. If subjects are asked to evaluate a statement that involves cheating on a social contract, they perform amazingly well. If the setting is different, they perform poorly. We have no general capability for doing logic. Formal capabilities are linked with particular situations.

Cosmides and Tooby argue that cheater detection is the pivotal issue in cooperation. Results of the Wason selection test are poor if the aim of testing a conditional thesis is to detect genuine altruism, in which a person benefits others while harming himself or herself. Cheater detection works much better. This fits well with evolutionary theory, which entails that genuine altruism should be rare, while cheating should be common.

Cosmides and Tooby's line of reasoning fits in with the predominant emphasis on self-interest in scientific and philosophical views of human behavior.

Their rash dismissal of the objection that real life is not like a prisoner's dilemma is inappropriate. Biology is rife with examples of cooperation even among animals, where benefits are mutual while costs are low or nonexistent (see section 6.5). Cosmides and Tooby pit cheating against genuine altruism in considering experiments on logical reasoning. They overlook forms of cooperation that do not allow of cheating (see examples in section 6.5). Many authors have indeed argued that cooperation with benefits for all concerned are common (e.g., see Boon and Holmes 1991; Frank 1988; Pope 1994; Ridley 1996).

The links between logically correct reasoning and cheating, however interesting, do not prove as much as Cosmides and Tooby think. Critical faculties drawing on logic are generally dormant. At times, we must be on the alert—for example, when we are cheated. Evidence showing that we do not always exercise critical faculties does not imply that such faculties are intrinsically connected with specific situations. We could as well speculate that nonspecific critical faculties exist, which are activated only under special conditions (for critical comments on the domain-specificity thesis, see Mithen 1996).

Many behaviors, on the face of it, are not linked to fitness. Some researchers like to collaborate with colleagues; others prefer to work on their own. These differences in research strategy do not affect fitness in the sense of reproductive survival. We could assume that behavior is, nonetheless, driven by a calculus of self-interest. That is a common assumption in many disciplines. But this assumption, which is problematic in itself, does not call for the evolutionary foundation that Cosmides and Tooby envisage.

A calculus of self-interest, as envisaged by Cosmides and Tooby, implies that we should often be dishonest when we can get away with it. However, it is impossible to change personality at will. Occasional "honesty" can only be a dilute form of genuine honesty. The chances are that others will recognize a lack of genuine honesty. That is not in our best interests. People who are genuinely honest may do much better. But these people would not usually be dishonest in situations where dishonesty appears to pay.

Frank (1990) has argued this point in a convincing way. One of his examples concerns tipping in a distant city. The motive of tipping in a distant city, according to Frank, is not to avoid being caught. "My failure to tip in the distant city will make it difficult to sustain the emotions that motivate me to behave honestly on other occasions. It is this change in my emotional makeup, not my failure to tip in itself, that other people may apprehend" (Frank 1990: 95).

Frank argues that according to standard models of game theory, genuine cooperation is unlikely to evolve in a population if cheaters and cooperators look alike. But if perfect recognition of cooperators and cheaters is possible, cooperation is the most viable strategy. Real populations allow for some recognition. Thus, seemingly irrational behavior may be rational in a larger framework. Frank's account explains unopportunistic behaviors that evolutionary accounts envisaged by Cosmides and Tooby fail to explain.

Cosmides and Tooby might object that tipping in a strange city yields benefits after all, by way of an emotional makeup that pays off in different situations. They might argue to enlarge the set of costs and benefits to accommodate anomalies. However, attempts to accommodate all apparent anomalies makes the thesis of universal self-interest that feeds cost–benefit approaches vacuous: people help strangers because it makes them feel good. People torture themselves because they are masochists, and torture gives them what they want. Any conceivable behavior can be construed as a source of benefit on a sufficiently inclusive notion of benefit. The presumption that self-interest is at the core of human behavior thus becomes a boring tautology (see section 6.2).

Cosmides and Tooby might reply that they would restrict the set of costs and benefits through evolutionary considerations. An apparent benefit is a genuine benefit only if it enhances fitness, however indirectly, and costs count only if they decrease fitness. At first sight, tipping in a strange city does not benefit me. If it does influence fitness, then its effect should be negative. But on second thought, it is beneficial because it results from an emotional disposition that pays off in different situations. Thus, it should enhance fitness, while self-torture, for example, should decrease fitness.

This argument invites the question why natural selection has failed to produce more flexible emotional dispositions. More benefits would accrue to me if I could feel like a genuine cheat in some situations, and like a genuinely honest person in other situations. Friends of evolutionary self-interest may respond that it is impossible for natural selection to do this. Whatever the selective regimes, elephants cannot evolve into winged creatures that catch bats up in the air, and we will not be blessed with personality switches. In brief, evolution by natural selection is constrained by features of organisms that are hard to change.

Whenever we come across a behavior that is anomalous from an evolutionary point of view, we can postulate constraints to make the anomaly disappear. We can stipulate that causal factors responsible for apparent anomalies are constraints. The stipulation may make sense if we are able to identify particular factors as constraints (see chapter 4). But Cosmides and Tooby have not produced plausible scenarios with concrete constraints.

The decision rules that Cosmides and Tooby appeal to are allegedly part of a cognitive program that human beings share. Psychologists may indeed show that our behavior fits in with some set of cost–benefit rules. But that does not imply that a cost–benefit program of which we are unaware is running inside of us. On touching a hot stove, a child rapidly withdraws its hand. That is a proper "decision" that benefits the child. But no withdrawal-decision rule that reckons with costs and benefits is implemented somewhere in the child's body. Although social interactions, unlike the withdrawal, are not generated by simple reflex arcs, the example suffices to illustrate that phenomena satisfying cost–benefit rules need not be produced by such rules. Rules envisaged by Cosmides and Tooby may be heuristically useful. They should not be reified into cognitive programs.

Evolution and Altruism

Cosmides and Tooby regard variability in behavior as a cultural overlay of more fundamental shared mechanisms explained by evolutionary theory. Their decision rules belong to such mechanisms. Common-sense experience suggests that people differ markedly in their willingness to cooperate—indeed, in all overt features that Cosmides and Tooby consider. This suggests that different persons follow different decision rules, if they follow rules at all. It is difficult to investigate causes of such differences in personality. But research indicates that genetic and environmental factors are implicated (Loehlin 1992). Cosmides and Tooby assume that according to evolutionary theory, core dispositions must be the same in all of us due to similarities in genetic makeup. That is false. D. S. Wilson (1994) has rightly argued that evolutionary thinking is compatible with the existence of adaptive genetic variation in personality. Phenomena such as genetic polymorphism, phenotypic plasticity, and ontogenetic shifts can account for this kind of variability. Moreover, some personality features may be adaptively neutral in that they have nothing to do with fitness as reproductive survival (Loehlin 1992).

The theory defended by Cosmides and Tooby is a variant of psychological egoism. It shares conceptual problems with other varieties of psychological egoism. Their evolutionary perspective does not solve these problems, because it distorts evolutionary biology. Cosmides and Tooby assume that genuine altruism is paradoxical, as it is incompatible with evolutionary thinking. The alleged paradox is analyzed in more detail by Sesardic (1995). I summarize his views in the next section.

6.7 THE PARADOX OF ALTRUISM: SESARDIC'S VIEW

Sesardic (1995) sets the stage for an analysis of the paradox of altruism by two notions of altruism: psychological altruism—altruism$_p$—and evolutionary altruism—altruism$_e$. He defines these notions as follows:

A is behaving altruistically$_p$ =$_{df}$ A is acting with an intention to advance the interest of others at the expense of his own interests.

A is behaving altruistically$_e$ =$_{df}$ the effect of A's behavior is an increase of fitness of some other organisms at the expense of its own fitness.

This distinction allows us to recognize that altruism$_p$ exists while granting that altruism$_e$ is evolutionary impossible. But an epistemic tension may still be generated by connections between the two species of altruism. Sesardic offers four apparently conflicting propositions as a reconstruction of the paradox of altruism:

1. Altruism$_e$ is a selectively disadvantageous trait.
2. Altruism$_p$ tends to lead to altruism$_e$.

3. Altruism$_p$ exists.
4. Altruism$_p$ is a product of natural selection.

Collectively, these propositions appear to imply that altruism$_p$ is both advantageous, because it is a product of selection, and disadvantageous, because it leads to altruism$_e$, which is itself disadvantageous. In principle, two strategies for dealing with the paradox are possible. The eliminativist strategy is to deny one of the propositions. It comes in four variants since each proposition could be denied. The reconciliationist strategy, which comes in three variants, aims to show that, contrary to appearances, the set of propositions is coherent.

Consider first the four variants of the eliminativist strategy as portrayed by Sesardic.

Regarding Proposition (1), we could argue that kin selection allows of altruism toward relatives. However, on a proper interpretation of kin selection, altruism toward relatives is not really altruism since it enhances fitness in the sense of inclusive fitness. The notion of inclusive fitness extends the notion of fitness as individual reproductive success by including reproduction of relatives as a function of genetic relatedness. The idea is that helping relatives fosters the propagation of our own genes. Reciprocal altruism, like altruism toward relatives, is not genuine altruism on either of the two definitions. A more promising way to challenge Proposition (1) is to appeal to group selection, which favors altruism even though it is individually disadvantageous. Sesardic holds that the power of group selection is weak, so that it does not constitute a reason to reject Proposition (1). He appeals here to a target article by Wilson and Sober (1994). These authors argue that the impact of group selection may have been underestimated. Sesardic interprets their article as implying that group selection is rare all the same.

Proposition (2) could be challenged in two ways. The link between the two kinds of altruism can break either because intended effects of human action do not in general correspond to actual effects, or because effects expressed in the currency of interests do not correspond to effects measured by fitness. The first possibility, according to Sesardic, is not realistic, since intentions to act are realized more often than not. Considering the second possibility, Sesardic notes that important interests are linked with fitness. Paradigmatic cases are protection of our health, avoidance of danger, and keeping our possessions. If we act against such interests, we naturally incur a loss of inclusive fitness. Thus, Proposition (2) appears to survive both challenges.

Proposition (3) might be denied on two grounds. First, some philosophers have argued that it is necessarily false since we are moved to action solely by the anticipation of pleasure. This thesis—psychological hedonism—has been refuted convincingly by many philosophers. Many actions can hardly be interpreted as a search for pleasure or self-gratification. Moreover, an act of, say, generosity would not give me the pleasant feeling of being generous if I undertook it solely for the sake of having this feeling. I would know in that case that I am not really generous. Second, someone could argue that the proposition is contingently false. Conceivably, apparently altruistic acts could result from hidden selfish motives to seek, for example, a reduction of negative arousal or mood enhancement. However, experimental psychologists have managed to distinguish between selfish and unselfish motivation, and their results disprove the egoistic hypothesis.

Proposition (4)—that altruism$_p$ is a product of natural selection—is likewise plausible according to Sesardic. Altruism$_p$ is a widespread feature sustained by powerful emotions that extend at least as far back as the period of hunter-gatherers. This is not a knockdown proof, but it suggests that altruism has roots in our evolutionary past.

So much for the eliminativist strategy. The reconciliationist strategy also comes in different variants. Sesardic recognizes three variants: the vestige theory, the byproduct theory, and the continuing adaptation theory.

According to the vestige theory, altruism is not now adaptive, but it has been adaptive under different conditions in the past. Sesardic knows of only one account that fits this theory—the kin selection account. The idea is that human beings in the past have lived in small groups consisting of close relatives. Thus, no need would have existed to distinguish between relatives and nonrelatives, and nondiscriminatory altruism would have worked well. Sesardic rejects this possibility for two reasons. First, our Pleistocene ancestors presumably lived in fairly large groups. Second, even if they had lived in smallish kin groups, distinguishing between degrees of relatedness would have been adaptive.

The byproduct theory contains the assumption that selection *of* altruism$_p$ has occurred without selection *for* altruism$_p$. Selection for another trait linked with altruism$_p$ may account for the prevalence of altruism$_p$. Sesardic considers two examples of theorizing along these lines. First, some philosophers have argued that selection has fostered human rationality, and that being rational entails being concerned for the interests of all rational beings. But this is no longer regarded as a powerful argument. The idea that unaided reason leads to substantive conclusions is philosophically suspect. Apart from this, social scientists would argue that purely rational considerations are motivationally inert. Moral attitudes cannot derive from rationality alone. Second, it is conceivable that natural selection promotes the trait of group conformity, and that altruism$_p$ is consequently favored by a process of cultural group selection. Sesardic argues that this theory, which is due to Boyd and Richerson (1985, 1990), may be plausible.

According to the continuing adaptation theory, altruism$_p$ is an adaptive product of natural selection. At first sight, this is odd in view of Propositions (1)–(4). Altruism$_p$ is supposed to lead to altruism$_e$, and altruism$_e$ is selectively disadvantageous. This suggests that altruism$_p$ could not be adaptive. Yet, Sesardic regards this as a fruitful idea. He appeals to prisoner's dilemma games in which each player has the choice between cooperating and defecting. The dilemma is that a person's payoff is maximal if that person defects while the opponent cooperates. The same goes for the opponent. Someone's payoff is minimal if that person cooperates while the opponent defects. The same is true for the opponent. It would be wise to cooperate in this situation, with intermediate payoffs for both players. Yet the two players will defect. Sesardic considers the following variant of the game. If someone has a disposition to cooperate, and the opponent has the first choice, and the opponent is able to recognize the person's disposition, then mutual cooperation will ensue. It would be more profitable for the person to defect, but that person's disposition ensures that the two players cooperate. Sesardic notes that some people would regard the disposition as an advantageous trait, so that we are dealing here with a denial of Proposition (1). He grants the point, but he sticks to his own interpretation, since the manifestations of the disposition decrease fitness.

All in all, Sesardic suggests that possibilities for altruism$_p$ in an evolutionary context are limited.

6.8 DISSOLVING THE PARADOX OF ALTRUISM

Sesardic's analyses have a plausible ring on the assumption that his key concepts are perspicuous. The entire conceptualization, however, is problematic. Let us consider how he conceptualizes potentially available options to get rid of the paradox of altruism.

Sesardic argues that Proposition (1) is presumably true, but it is actually useless. The proposition presupposes that it makes sense to distuinguish between selectively advantageous traits and selectively disadvantageous traits. But such a general distinction is unworkable because "advantage" is a relative notion. Traits should be conceptualized as advantageous relative to alternative traits, against a background of an array of other traits and a particular environment. Thus, relative advantages depend on the comparison we wish to make. Without a detailed specification of context that privileges particular comparisons, we would be left with the trivial thesis that features of organisms are always advantageous in some respects and disadvantageous in other respects. Being-advantageous-in-general is a useless notion (see chapters 3 and 4). The same may be said for other crucial concepts used by Sesardic. "Fitness," for example, is notoriously ambiguous (see chapter 2). Furthermore, "inclusive fitness" is a problematic extension of the concept of fitness (Sterelny and Kitcher 1988).

Sesardic relies on general placeholder concepts that need to be replaced by more specific notions. He is not aware that evolutionary accounts call for natural history.

The most salient point made by Sesardic concerns group selection. He maintains that Wilson and Sober suggest that individual selection mostly overrides group selection. In fact, they argue against this view. They also show that common demarcations between kin selection and group selection may be conceptually inappropriate. Their article is followed by an extensive peer review that should have a sobering effect on anyone who would like to defend general theses about selection. Different authors use central notions such as "adaptation," "fitness," "kin selection," and "group selection" in different senses. This indicates that we need a much richer array of concepts to cover evolutionary thinking. The meaning of common key notions is context-dependent. Once we start specifying contexts, natural history takes the place of sweeping generalities that dominate much thinking about evolution outside biology.

Sesardic envisages two ways to challenge Proposition (2). First, the link between the two kinds of altruism may break because intended effects of acts need not in general correspond with actual effects. Against this, he argues that intended effects are realized more often than not. That may be true, but we still have to envisage unintended side effects. Sesardic only considers such effects in a footnote, where he grants that altruism$_p$ may not lead to altruism$_e$ if acts, unknown to the subject, have fitness-increasing consequences as unintended effects (Sesardic 1995: 135). Should we regard this possibility as an example of altruism$_p$ as defined by Sesardic? His view on this matter is not entirely clear. We have to decide if altruism$_p$ must be defined with reference to interests-as-intended only. If we so define it, then we need additional concepts to cover unintended effects. Sesardic's analysis would then be misleading, as Proposition (2) is unduly restrictive. Alternatively, if we opt for a less narrow definition of altruism$_p$, then Sesardic's defense of Proposition (2) is not generally valid.

Second, the link between the two kinds of altruism may break because effects expressed as interests do not correspond to effects measured by fitness. Sesardic rejects this possibility because important interests such as health, avoidance of danger, and keeping possessions are allegedly linked with fitness. He envisages relatively extreme cases of altruism$_p$ that may involve a loss of fitness, but may well be rare (Sober 1991). We should note that more representative, less extreme cases need not entail a loss of fitness.

The problem of altruism gets more elusive as we enlarge the category of items that count as interests. If interests are defined broadly, without a specification of currency, and if the definition of altruism$_p$ is construed to cover unintended effects in addition to intended effects, then psychological egoism does become tautologically true and psychological altruism becomes a contradiction in terms. Barring the exception of compulsion, we do the things we desire to do. Add the premise that it is in our interest to have our desires satisfied, and we get the conclusion that all our acts are self-serving.

This would amount to a denial of Proposition (3), that altruism$_p$ exists. The argument suggests that psychological egoism is necessarily true. Sesardic rejects this possibility. He argues that the main support for psychological egoism comes from hedonism—the view that the expectation of pleasure is the sole motivation of human action—and that hedonism is empirically false. This argument is unconvincing, because hedonism focuses on a restricted set of interests.

Literature in ethics about the issue is confusing. Regarding the merits of psychological egoism, and the allied normative thesis of ethical egoism, four options are conceivable: that it is contingently true, contingently false, necessarily true, or necessarily false. All the options have been defended in the literature (see section 6.3). This is possible because different authors specify the general, abstract notions of altruism and egoism in different ways. They depart, for example, from different categorizations of costs and benefits.

Suppose that a friend helps you with your homework, and you help her in return with the dishes; her helping you is part of reciprocal altruism. Reciprocal altruism is not real altruism, according to Sesardic. Is your friend an altruist if you do not help her in return? Perhaps not. If you like her for her help, she does get something in return. Surely your liking her is in her interest. Thus, her helping you may represent psychological egoism (E_p) in Sesardic's sense. But perhaps the wish to be liked was not her motive. Then it is unclear if we are dealing with E_p; we have to decide here about the exclusion or inclusion of unintended benefits in the definition of E_p. If you do not like your friend for her help, her act does not, on the face of it, represent E_p. But we could still argue that she may feel good about her act since it fits in with her conception of morality. Thus, we could refuse to see her as an altruist since her feeling good is in her interest. If this interest is not what motivates her, we have to decide again about the exclusion or inclusion of unintended benefits in the definition of E_p before we can attach the label of E_p or non-E_p to the act.

Considering which things should count as benefits, we may take a lenient stance. That would allow us to construe even the most outrageous behaviors as a matter of self-interest. If your friend wants to help you with the dishes, and you refuse her help, she may feel so disappointed that she commits suicide. Her overall interests may not be served by this, but she benefits in that her feeling of disappointment disappears. Indeed, she might regard suicide as matching her overall interests if being rejected by you of all persons feels for her like the worst thing that could ever happen.

Sesardic regards Proposition (3) as a true empirical thesis. My analysis indicates that the thesis, like Proposition (1), is useless since terms such as "egoism" and "altruism" are overly general and vague placeholder concepts. They represent a heterogeneous array of concepts. We need more specific concepts that fit particular contexts. Our conclusion should again be that general theorizing as pursued by Sesardic needs to be replaced by natural history.

Sesardic also discusses another possibility to deny Proposition (3). Apparently altruistic acts could derive from hidden selfish motives. We may simply note that hidden or unconscious motives constitute a controversial subject of psychology. Sesardic maintains that psychologists have disproved a variant of the egoism thesis that relies on hidden motivation. That is an unwarranted simplification of theorizing in psychology.

Proposition (4), together with the other propositions, engenders the paradox of altruism on the assumption that products of natural selection are advantageous. Sesardic should have listed this assumption as a separate proposition. The assumption is false. In comments on Proposition (1), we have already seen that the general notion of disadvantage is unclear, such that (1) is useless. The same goes for the assumption. But we can work out the assumption for particular contexts, and consider specific variants. The most relevant context would be population genetics—according to many researchers, the core of evolutionary biology. The idea that selection produces advantageous traits, in this context, may be taken to mean that selection maximizes fitness in the sense of expected reproductive survival. This is false as a general thesis. Any textbook of population genetics offers models for particular situations in which fitness is not maximized (Hartl and Clark 1989).

Sesardic's dismissal of the vestige theory is also simplistic. Biologists would agree that features of organisms may represent adaptations to conditions that existed in the past. To investigate this, they would resort to natural history in the form of phylogeny analysis (Brooks and McLennan 1991). Considering human behavior, Sesardic, like many other authors, mentions the Pleistocene as a significant period of geology. Biologists concerned with phylogeny would note that organisms typically exhibit combinations of traits that emerged in different periods of history. This should apply to human behavior also. Unfortunately, it is hard to reconstruct the phylogeny of behavioral traits, since behavior does not fossilize. Therefore, sound empirical theses about the evolutionary history of human behavior are hard to come by. Speculations about human beings in the Pleistocene are interesting, but they amount to mere speculation about a restricted period of significant history.

Regarding the byproduct theory, we should consider a possibility disregarded by Sesardic. It is conceivable that altruism toward kin and reciprocal altruism, neither of which counts as genuine altruism$_p$, would be fostered by a general disposition to help persons we are in close contact with and feel empathetic toward. It may be difficult to develop a disposition that is more discriminate. So a genuine form of altruism$_p$, admittedly not fully indiscriminate altruism$_p$, may arise as a byproduct of kin selection and selection for reciprocal altruism.

Sesardic dismisses another variant of the byproduct theory which assumes that it is rational to be moral. He suggests that this is false. But it is unclear since the term "rationality" stands for many different concepts. Ethicists have written much about the subject without reaching consensus. Disputes about it are in the same logical boat as the disputes over egoism.

Considering the theory of cultural group selection, I would side with Sesardic. The theory may help to explain altruism, but to decide if it is viable we need more details.

The most serious explanation of altruism$_p$, according to Sesardic, is by way of the continuing adaptation theory, which relies on a particular version of the prisoner's dilemma. But this theory is not very satisfactory. The envisaged disposition to cooperate is advantageous because without it outcomes would be worse. Sesardic argues that it would be even more advantageous to have the disposition without manifesting it, but that is an odd argument. Dispositions that are not manifested under the proper conditions are not really dispositions. Considering the example, we should instead say that particular outcomes cannot be realized due to constraints on behavior. Within the space of realizable outcomes, the two parties choose the best option. We are dealing here not with altruism, but with mutualism.

We can but conclude that evolutionary biology is compatible with a great variety of behaviors, including various forms of altruism. No paradox of altruism exists.

6.9 CONCLUSIONS

Conceptual analysis indicates that the search for highly general theories in ethics had better be curtailed. The doctrine of ethical egoism is a telling example. It keeps generating controversies due to conceptual rather than substantive problems. The controversies are sterile since crucial general terms such as "altruism" and "egoism" are placeholders for heterogeneous categories of behavior. They do not help us formulate fruitful generalizations.

The analyses in this chapter show that selfishness is emphasized in unreasonable ways in science and ethics alike. Many researchers have argued that evolutionary biology implies that human behavior is ultimately selfish. The reliance on evolutionary biology is odd. The study of human beings now existing should be the primary source of evidence. Data concerning behavior in the remote past are hard to obtain. For that reason alone, it is implausible that evolutionary thinking could overrule evidence now accessible to common sense.

Human behavior is variable. Human beings can perform atrocious acts, and they can also exhibit genuine care for others. The assumption that all behaviors derive from universal dispositions should be distrusted. Evolutionary psychologists assume that such dispositions exist, and they see support in evolutionary biology. They argue that cooperation among human beings results from selection processes that enhance the fitness of individuals. Cooperation, from this point of view, would be a matter of self-interest, making genuine altruism impossible.

For conceptual reasons alone, the implied view of human nature should be rejected. Concepts such as "altruism," "egoism," and "cooperation" are ambiguous. The thesis of "psychological egoism," which construes all behavior as ultimately selfish, is vague. We should replace it by more specific theses representing natural history that allow for different forms of altruism, egoism, and cooperation.

Fashionable theories of evolutionary biology such as kin selection and reciprocal altruism appear to unmask altruism and cooperation as ultimately selfish. But these theories do not exhaust the explanatory versatility of biology. Animals show all sorts of cooperative behaviors. Some cooperative behaviors should be explained without recourse to fashionable theories. Alternative explanations exist. Evolutionary psychologists disregard these alternatives. They draw on a simplistic, biased version of biology that fosters unacceptable views of human nature.

It is true that evolution constrains behavior so as to preclude particular forms of altruism. Animals do not normally engage in behaviors that decrease their own fitness in the sense of reproductive success. Considering human beings, we should realize that this is a weak constraint. We do not usually help others if it precludes our getting children. Nor would anybody in his or her right mind require us to do that. If that is the message of evolutionary biology, we can happily take it for granted. The issue would seldom arise in daily life.

Evolutionary theory generates a paradox of altruism only if particular varieties of altruism exist that decrease the fitness of the altruist. Such varieties are rare. Sesardic holds that they are common since important interests such as protection of our health, avoidance of danger, and keeping our possessions are linked with fitness. His view is implausible. Most human beings would attend to their own survival. They would not give all their food to the needy and perish themselves from starvation. Nor would they give away all their possessions if that would preclude their getting and raising children. Few of us would take the moral stance that such extremes of altruism should be promoted.

Sesardic defends the common view that reciprocity should explain the paradox of altruism. The most obvious form of reciprocity concerns persons who help others and are then helped by them. But advantages may also take different forms, as in indirect reciprocity, in which secondary advantages for the altruist are due to third parties knowing about the altruism. Such indirect advantages may help to explain exceptional behaviors. Human beings may risk their own lives— for example, to save others from a burning house. But even such behaviors may have advantages for persons exhibiting them—these persons may earn a good reputation with other advantages in the wake of it.

Reciprocal altruism, broadly conceived, need not imply that altruists are motivated by advantages that their behavior has for themselves. Selfish motivation tends to backfire, and commitment without selfish motives may foster cooperation, with advantages for all concerned. If we have a caring attitude toward other persons, then people will recognize this. The ensuing secondary advantages for ourselves do not imply that we are selfish after all. Genuine care for others fostering cooperation is more properly labeled altruism.

Altruism outside the framework of common sociality does exist, and some instances of it may be difficult to explain. But we should not overvalue it. If a friend continually tried to help you while opposing every attempt on your part to reciprocate, you would presumably be irritated and might even consider terminating the friendship.

Biology is relevant for understanding issues such as altruism, egoism, and mutualism not least because it is the science that best confronts us with the phenomenon of variation. Human behaviors, like all behaviors of organisms, exhibit variation to an astonishing degree. Social interactions come in many different kinds. Highly general theories that rely on a single notion of altruism and a single notion of egoism are a poor means to understand these interactions.

7

Evolution and Culture

7.1 INTRODUCTION

Evolutionary theory applies to all organisms, and man should not be an exception. However, in many respects man is special. We have a sophisticated language, our consciousness takes the form of self-awareness, and we are unsurpassed in creating cultural artifacts. One wonders if evolutionary theory can explain these things. It is reasonable to speculate that the theory explains at least some aspects of capabilities that characterize man. Our capabilities depend on special features of our brains, and natural selection should have played a role in the origin and maintenance of these features. But the theory of biological evolution cannot explain everything about us. For one thing, the huge differences that exist among human cultures could not be explained by diversifying natural selection acting on genes. Biological evolution does not account for the rapidity of cultural change.

Yet this does not imply that evolutionary accounts are impossible. We may opt for broad definitions of "evolution" and "selection" that leave room for non-biological forms of evolution and selection. Just as natural selection involves the differential replication of genes, different forms of selection may involve the differential replication of other entities.

Many scholars have conceptualized cultural selection in addition to natural selection. I summarize here the views of Durham (1991), whose account is comprehensive and detailed. Other well-known accounts of evolution and culture (Boyd and Richerson 1985; Cavalli-Sforza and Feldman 1981; Lumsden and Wilson 1981) cover only part of the territory charted by Durham. I also comment briefly on a more speculative recent theory of Blackmore (1999). Next, I discuss an example of my own to defend the thesis that phenomena of cultural "selection" are so complex and diverse that attempts to elaborate a single, general theory of cultural evolution are futile. Cultural evolution calls for a natural history approach. Lastly, I comment on the views of evolutionary psychologists on the evolutionary origin of our capacity for culture.

7.2 NATURAL AND CULTURAL SELECTION

Durham (1991) departs from the reasonable premise that genes and culture constitute two distinct but interacting systems of information inheritance within human populations. He recognizes five system requirements for evolutionary change. (i) Both systems can be divided into recognizable subunits of transmission and inheritance. (ii) Within all populations there are sources of variation in these units which create alternative forms at least occasionally. (iii) One or more mechanisms of transmission exist through which these units are conveyed among the individuals of a population. (iv) Natural selection is the main but not the exclusive means of modification in organic evolution. Counterparts exist in cultural evolution. (v) The necessary sources of isolation exist such that diversification can occur.

Considering units of transmission (informational entities), Durham notes that these vary widely with respect to scale and complexity. Examples of cultural units are morphemes of language, complex ideas, beliefs and values, entire languages, ideologies, symbol systems, and culture pools. In a given context, the answer to "What is the relevant unit of transmission?" is essentially an empirical one: whatever is both variable and differentially transmitted, regardless of its size and complexity. Durham adopts a term coined by Dawkins (1976), "meme," for cultural units. Memes, like genes, affect the fate of their carriers—they have "ecological consequences." Also, memes, like genes, come in varieties. Different varieties of a gene are called alleles; different varieties of memes are called allomemes. Additional analogies are easily construed. Thus we have the analog of biological fitness—the cultural fitness of an allomeme—which is its relative rate of social transmission per unit of time.

Sources of variation and mechanisms of transmission are not the same in the two types of evolution, but their roles are analogous. For example, innovation as a source of variation in cultural evolution functions like mutation in biological evolution.

Durham distinguishes five different modes of gene–culture relationship: two interactive modes, in which a change in one system causes a change in the other system, and three comparative modes, in which changes are independent.

1. *Genic mediation* is the first interactive mode; in this, genetic changes cause changes in primary values—that is, values deriving from biological features—and thereby cultural evolution. The classification of referents of color terms, which depends on genetically determined neural structures and functions in-

volved in color vision, is an example. This mode is prominent in the theory of Lumsden and Wilson (1981).

2. *Cultural mediation* is the second interactive mode; in this, cultural change affects genetic change. Sickle-cell anemia is an example. Persons with this disease have an abnormal form of hemoglobin, causing serious anemia. Persons who are homozygous for the gene involved die before they are able to reproduce. Heterozygous persons are also afflicted, but they are able to reproduce. The disease is especially common in various regions in Africa. It is so crippling that it seems odd that natural selection has failed to eliminate it or at least to decrease its prevalence. The explanation is that persons who are heterozygous for the sickle-cell gene have a higher resistance to malaria. This causes the disease to persist in regions where malaria is endemic. Now, the incidence of malaria is affected by cultural factors. Differences in agriculture cause differences in densities of mosquitoes, the vectors of the malaria parasite, and thereby differences in the prevalence of anemia. Yam-based agriculture causes higher mosquito densities. Thus, changes in agriculture, which are a matter of cultural evolution, have a marked effect on natural selection in this case.

    Lactose absorption is another example. Adult human beings originally were unable to absorb lactose present in cow's milk. In some populations, cultural changes fostering dairying and milk use led to a process of natural selection enhancing the capability of absorbing lactose. As a result, genetic differences now exist among populations in this capability.

3. *Enhancement* is the first comparative mode; in this, secondary values—that is, values due to collective experience and social history—influence social transmission in such a way that differences in cultural fitness exaggerate differences in biological fitness. The monomarital principle in Tibet is an example. According to this principle, in each generation of a family only one marriage can be contracted. All sorts of matrimony are compatible with the principle. Thus, we find in Tibet various forms of polyandry and polygyny. Perceived consequences concerning stable land tenure and resource availability are the cultural driving force underlying the monomarital principle. The cultural decisions concerning matrimony have clear inclusive fitness advantages. Incest taboos are also an example of enhancement. The taboos are transmitted through cultural selection, but they affect biological fitness.

4. *Neutrality* is the second comparative mode; in this, cultural selection favors adaptively neutral allomemes. Differences in the form of tools, and in linguistic symbols, are obvious examples.

5. *Opposition* is the third comparative mode; in this differential transmission of allomemes decreases the inclusive fitness and adaptedness of members of a population. Socially imposed land-tenure constraints are an example.

Cultural systems tend to evolve toward relationships of enhancement for those who have the power to shape them. Cultural selection generally favors allomemes that improve, or at least do not diminish, the reproductive fitness of their selectors under current conditions.

Blackmore (1999) has recently developed a theory that emphasizes, more so than Durham's theory, that cultural selection has to a large extent become an independent force. She holds that the ability to imitate, which is rare in animals, has been the driving force of cultural change in man. According to her, the social skills that others have singled out as directly responsible for our large brain were in fact responsible for the prior step of acquiring imitation, which ensured the spread of memes. As soon as our ancestors crossed the threshold into true imitation, a second replicator was unwittingly unleashed. Only then did the "memetic" pressure for increasing brain size begin (p. 76). This is an intriguing speculation.

I regard examples offered by Blackmore as less than compelling. Here is what she has to say about the tendency of humans to talk a lot.

> ... let's compare two types of meme. Suppose there are instructions encouraging people to talk a lot. These might come in many forms, such as embarrassment at being silent in company, or rules about making polite conversation or entertaining people with chat. Now suppose there are other memes for keeping silent, such as the suggestion that idle chat is pointless, a rule of quiet etiquette, or a spiritual belief in the value of silence. Which will do better? I suggest the first type will. People who hold these memes will talk more; therefore, the things they say will be heard more often and have more chances of being picked up by other people. [Blackmore 1999: 85]

Frankly, I do not like garrulous persons. And I have friends who manage without talking to get the message across that being silent can be agreeable. It is true that the things said by people who talk have more chances to be picked up by other people than things left unsaid by silent people. But it does not follow that this suffices to induce talking in the others. *If* the others talk, they may repeat what they have heard. But it does not follow that they will like to talk.

Here is another example of a "memetic" explanation offered by Blackmore. She considers that "memetics" is a good candidate for explaining altruism in man:

> ... if people are altruistic they become popular, because they are popular they are copied, and because they are copied their memes spread more widely than the memes of not-so-altruistic people, *including the altruism memes themselves*. This provides a mechanism for spreading altruistic behavior. [Blackmore 1999: 155]

What about rape and murder as common phenomena of war? Are soldiers who commit such atrocities more popular with their comrades? Is that why the behavior spreads? If so, then "memetics" explains altruism and cruelty alike, which is suspect. Blackmore's explanation would be nice if altruism were a predominant mode of human behavior. It is not. We have to find out what circumstances promote particular forms of behavior. That calls for natural history. Sweeping claims like the ones endorsed by Blackmore are to be distrusted.

Evolution and Culture 103

The examples of explanations that Durham offers are more convincing, as he provides the requisite details.

7.3 AGAINST OVERARCHING THEORIES OF CULTURE

Durham's examples are well documented, and I regard them as convincing. Particularly valuable is his evidence that all sorts of relations exist between cultural change and genetic change. But I feel uneasy about the idea that we could develop a general theory of cultural change or, more particularly, cultural selection. Units and mechanisms of cultural change are so diverse that it is futile to force them into the straitjacket of a single theoretical framework. Durham, for that matter, grants that this diversity does exist. I work out an example of my own—drug therapies versus diet therapies in rheumatoid arthritis—to illustrate this.

NSAIDs (nonsteroid antiinflammatory drugs) are used on a massive scale to palliate symptoms in inflammatory diseases such as rheumatoid arthritis. But as many studies have shown, the drugs have dangerous gastrointestinal side-effects that frequently result in death (e.g., see Griffin 1998; Wolfe 1996). The drugs keep being used despite the existence of less harmful alternative treatments. For example, particular diets reduce disease activity in rheumatoid arthritis (Ariza-Ariza, Mestanza-Peralta, and Cardiel 1998; De Luca, Rothman, and Zurier 1995; James and Cleland 1997; Kjeldsen-Kragh et al. 1995; Mera 1994; Rothman, De Luca, and Zurier 1995).

Why are NSAIDs still used on a grand scale even though diet therapies may represent a better option? Or, in the terminology of this chapter, what mechanisms provide NSAIDs with a pronounced selective advantage over diets? Let me review some potential sources of bias.

Abraham (1995) demonstrates that research by the pharmaceutical industry has been biased in favor NSAIDs. That should not surprise us. What is surprising is that incontrovertible evidence exists that authorities responsible for regulation have, on a grand scale, acted in violation of their own guidelines in allowing NSAIDs to be marketed.

We can describe this situation also in the terminology of cultural selection. Selective bias in ideas promulgated by the industry results from a mechanism involving competing motives—the profit motive, and more noble motives concerning the well-being of patients. At times, perhaps frequently, the profit motive overrules the more noble motives. In addition to this, governmental agencies suffer from a selective bias in favor of the industry. I conjecture that these forms of bias result, to a large extent, from conscious deliberation. It is implausible, for example, that governmental agencies could act against their own guidelines without noticing this.

Bias in favor of the industry may also exist in medical research itself. As far as I know, NSAIDs have not been investigated in this respect, but convincing data are available for calcium-channel antagonists. Stelfox, Chua, O'Rourke, and Detsky (1998) have shown that the chances that researchers observe positive effects of a drug are way above average if they are paid by the industry manufacturing the drug. The industry is apparently a selective agent causing bias in scientific research. Considering this form of bias, I would conjecture that to a large extent it is subconscious. The alternative would be to assume that deliberate fraud is extremely common in science.

Medical research also exhibits bias in the choice of subjects to be investigated. At the time of writing (January 1999), Medline lists 7,573 articles on drug therapy, as against 99 articles on diet therapy. This fits in with a pervasive negligence of ecology in medical research (Chivian, McCally, Hu, and Haines 1993; Garrett 1994; van der Steen 1998). The emphasis on drugs in research is again explained, in part, by the pharmaceutical industry acting as a selective agent. But the negligence of ecology also has other sources. Medical science had assimilated much biology prior to the advent of ecology. It may well have taken a self-perpetuating shape that worked against the assimilation of ecology. In any case, university students of medicine are hardly trained in ecology. Thus, cultural selection by way of the transmission of knowledge in higher education is yet another factor helping NSAIDs win from diet therapies.

We should add to this that modern techniques of food processing cause the destruction of natural components of diets—not least, unsaturated fatty acids, including essential ones. This fosters degenerative and systemic diseases (Erasmus 1995), and it may well contribute to etiology in rheumatoid arthritis. Selection processes fostering modern food-processing have therefore contributed to the use of NSAIDs.

A single discipline or interdiscipline covering all the factors promoting the use of NSAIDs does not exist. Nor is it likely that any such discipline or interdiscipline will be created. There is no need for this. To assess the use of NSAIDs, we can rely on existing disciplines (medicine, biology, sociology of politics, sociology of science, history of science, methodology, to name a few) and other sources of knowledge, but we need not integrate them to understand what is happening. The operative factors come in different kinds which have little in common apart from their effect on the use of NSAIDs. We may choose to say that all the factors exert their influence by processes of cultural selection, but that does not help us understand much, because the label of selection thus represents utterly different things. Medical students are taught to focus on genetics, biochemistry, and anatomy and to disregard ecology. The pharmaceutical industry promotes the use of drugs rather than the use of diets. The food industry saddles us with foods that promote particular diseases and thereby the use of drugs. It is easy to understand these things, and calling all of them by the name of cultural selection does not lead to new insights.

Natural selection processes, in the ordinary sense, are a different matter. True, these processes are also diverse, but they have much more in common than cultural selection processes. Considering natural selection, I have argued in previous chapters that the diversity of factors involved precludes the elaboration of overarching general theories. We have to be content with a coherent body of natural history. Population genetics—according to many biologists, the core of evolutionary biology—comes closest to generality. But its models cover only a particular aspect of selection processes. The common denominator here is genetic change in populations in which some alleles are replaced by others. The agents of selection (if present) are not specified; they are covered by selection coefficients. A full-fledged explanation of a particular selection process will have to specify selective agents. Thus, we have to enrich population genetics with ecology. The result is natural history because the diversity of selective agents does not allow of generalities. Whatever generality there is resides in the effects. So we have diversity of causes and commonality of effects in the form of genetic change. In cultural selection, causes are much more diverse, and effects are heterogeneous as well. That precludes even the development of a coherent body of natural history. Patterns of road building, recipes for cooking, and the conduct of religious services have little in common. Let us not try, then, to cover them by a single body of knowledge.

Let me repeat that, in spite of all this, I regard Durham's view as valuable and insightful. The existence of various relations between cultural change and biological evolution is interesting. The relations reveal to what extent evolutionary biology can explain human behavior. But we should keep in mind that cultural change is not a unitary phenomenon, and that it is futile to aim at a unitary theory of relations between biology and culture.

7.4 EVOLUTIONARY PSYCHOLOGISTS ON CULTURE

Authors such as Durham, investigating interactions between natural selection and cultural selection, mostly take the existence of culture in the human species as a point of departure. Other authors have made the evolutionary origin of culture their subject of research. In evolutionary psychology, the focus is mainly on this subject. Broadly speaking, the goal of evolutionary psychology is the explanation of psychological processes as biological adaptations. In chapter 6, I analyzed the views of evolutionary psychologists on the more specific theme of altruism and egoism. Let me turn to more general views that they defend.

Evolutionary psychologists (Buss 1991, 1995, 1999; Cosmides 1989; Cosmides and Tooby 1994a, 1994b, 1995; Tooby and Cosmides 1992, 1995) oppose the so-called standard social science model (SSSM), which assumes that human nature is infinitely malleable through culture, and that it has no fixed, hard-wired capacities. Instead, they postulate an evolved, universal human nature, impenetrable for cultural influences, and they regard existing variability as ephemeral. The SSSM considers cultural influences the sole explanation of human behavior; evolutionary psychologists counter that culture is not something that causes or explains anything, but rather is in need of explanation itself. The existence of universal hard-wired evolved psychological mechanisms does not mean that all behavior is innate. Evolved mechanisms can produce varied patterns of behavior as a result of different inputs.

Considering mind as a biological organ suggests a second thesis contrary to the SSSM. Mental mechanisms must be knowledge-rich rather than general content-free reasoning capacities. This is because they have evolved as adaptations for a specific kind of environment and therefore embody knowledge of that environment.

Evolutionary psychologists regard the Pleistocene as the period in which the human mind took shape. The human environment of selection posed a number of unrelated standard problems (cheater detection, memory for places, stereo vision) in the ancestral environment that were unique, and no general solution existed for all of them. This implies that the mind must be a collection of cognitive tools, with different applications for different situations. Each contains a dedicated data base; mind is something like a Swiss army knife of unrelated tools dedicated to specific tasks, rather than a universal general intelligence.

Evolutionary psychologists assume that adaptations are recognized by their complexity. Adaptations are too well organized to be the product of chance. Presumably the formation of complex capacities takes too long to have changed over the past few thousand years. Since mental capacities are complex adaptations, the basic make-up of the mind must have evolved long before modern society took shape.

The undertaking of evolutionary psychology is interesting and intriguing, but I do not believe that it has delivered any of the goods promised (for a more detailed survey, see Looren de Jong and van der Steen 1998). When evolutionary psychologists postulate mental modules that generate engineering solutions to adaptive problems that our ancestors faced, they are insufficiently aware of the need to flesh out biological details by way of a natural history approach. This is best brought out by the sophisticated analysis of adaptation explanations by Brandon (1990).

Brandon formulated five conditions for legitimate and complete adaptation explanations. First, evidence must exist that selection has occurred. Second, we need to specify why selection has occurred, by citing some ecological factor that was effective in selection. Third, the trait must be demonstrably heritable. Measuring heritability requires variation; however, selection destroys the variation on which it acts. Therefore, it may be difficult to demonstrate that the less adaptive traits have existed, but disappeared. Fourth, we need information about population structure regarding gene flow, and about the structure of the selective environment. Finally, we have to know about phylogenetically primitive traits, and derived traits. The upshot of these conditions is to anchor the notion of adaptedness firmly in causal history. Adaptedness (in the sense relevant here) is a historical notion, and adaptive explanations require historical information showing under what conditions (regarding extent of genetic variation, selective forces in the environment, the degree of heritability) selection has actually occurred (see also Griffiths 1996b). Without this information, adaptive explanations tend to take the form of just-so stories, which are easy to generate but difficult to test (Gould and Lewontin 1979). Brandon's example of an adaptive explanation is heavy-metal tolerance in plants. He shows that this example fits the five requirements reasonably well. It is exceptional in this respect. Seldom are we able to spell out adaptive explanations in detail.

Richardson (1996) has applied Brandon's five criteria to claims of evolutionary psychologists that rationality, language, and capacities for social exchange have been selected as adaptations. His conclusion is that all proposals fail dismally to provide evidence about the actual history of selection. With respect to rationality, language, or social skills nothing is known about the variation in such traits between groups, the degree of selective advantage, the ecological mechanisms producing selection, the degree of heritability, population structure, or evolutionary precursors. This should come as no surprise, because behavior does not fossilize.

Also problematic is the focus of evolutionary psychology on Pleistocene conditions. Biologists would argue that traits have accrued over long stretches of the evolutionary tree, and they consider the analysis of phylogeny as an important empirical task. It is odd to restrict phylogenetic analysis to just one historical period, as Cosmides and Tooby do. Biologists have well-established formal methods to assess the origin of traits during phylogeny (e.g., see Brooks and McLennan 1991, who consider several types of adaptation scenarios which account for ecological and historical information). Cosmides and Tooby's emphasis on the Pleistocene as the sole source of cognitive innovation does not sit well with approaches in live biology. They do not even seem to realize that phylogeny analysis exists as an important subdiscipline of biology.

Apart from this, there is no warrant for Cosmides and Tooby's assumption that, for lack of time, no substantive changes in complex adaptive features can have occurred since the Pleistocene. D. S. Wilson (1994) provides examples showing that rates of adaptive evolution can be much faster than evolutionary psychologists assume.

Next, the emphasis on universal traits and the depreciation of variability in evolutionary psychology does not sit well with evolutionary biology. Evolutionary biologists have argued for decades that individual differences in a population can be maintained by natural selection (Lewontin 1974; D. S. Wilson 1994). For example, frequency-dependent selection, where fitness depends on the mix of genotypes in a population, allows for variation in a population. D. S. Wilson (1994) offers a model of adaptive variation that includes both phenotypic plasticity—phenotypic variation not due to genetic variation—and genetic polymorphism—variation representing genetic differences. In either case, the environment may play important roles in the genesis of variation. Thus, individual differences may represent adaptive strategies in an environment, with different niches favoring different behaviors.

Buss (1999), in a recent book, tends to disregard variability or to explain it away in an ad hoc fashion. Let us consider some examples. He argues, for example, that regarding long-term mating, most women have a preference for tall, strong, athletic men with a lot of resources. Such features would ensure their and their children's well-being. This invites the question of how to explain the huge variability in physical strength existing nowadays among men, and likewise for physical attractiveness in women, allegedly the most important indicator of their mate value. Buss does note that evolutionary psychologists have undervalued variability. Considering mate preferences, he offers the following explanation of existing variability:

At a psychological level humans might have evolved mechanisms designed to assess their mate value relative to the individuals in their environment. Ancestral environments were probably populated with relatively small groups of people containing around fifty to one hundred individuals. . . . Assessments of relative mate value were probably fairly accurate. One result of those accurate assessments might have been to focus individuals' attraction tactics on potential mates within their own mate value range. [Buss 1999: 400–401]

Buss subsequently notes that in our culture, the media may present an unprecedented comparison standard, as they focus on extremely attractive women. Hence, intrasexual competition among women increases so as to provoke, in extreme cases, maladaptive body-image disorders such as anorexia and bulimia.

Against this explanation, I would argue that persons with a low mate value who opt for mates within their own mate-value range should on average have fewer descendants than persons with a high mate value. If features associated with mate value are heritable, as they should be on the basis of the proposed explanation, natural selection should have eliminated variability in these features. Thus, the scenario described by Buss does not really explain current variability. Neither does it explain the body-image disorders, which obviously do not increase mate value.

Another example concerns gang warfare in adolescents, which is common across America (Buss 1999: 304). Buss approvingly draws on research suggesting that the phenomenon allows of an evolutionary explanation. Gang members typically have more sex partners than nonmembers. Hence, gang warfare increases the fitness of participants and it should be promoted by natural selection. However, most adolescents do not join gangs. Again, we are saddled with unexplained variability.

Evolutionary psychologists are also not sufficiently aware of two other caveats concerning adaptationist analyses that are common knowledge in evolutionary biology. First, we should distinguish between effects of a feature that have been selected *for* and effects that have been selected *of* (Sober 1984). The heart has the effect of circulating the blood. This effect has presumably been selected for. Other effects such as the generation of heart tones are produced in the wake of this. Selection of heart tones has occurred, but it would be odd to postulate selection for heart tones. Second, Gould and Vbra (1982) have made the useful distinction between adaptations and exaptations. Adaptations are the result of selection. Exaptations represent new functions of features that were selected for different functions. Thus, we cannot infer from the existence of psychological functions in the present that these functions originated as evolutionary functions in some ancestral environment. Such inferences are possible only on the basis of knowledge of ancestral conditions, but this knowledge is painfully missing in the case of psychological capacities. Mere complexity is not a good pointer, since components of complex capacities may well be exaptations or byproducts of other features that have been selected for (Davies 1996).

To some extent, evolutionary psychologists are aware of this problem. Thus, Cosmides and Tooby (1994b: 95) note that activities we can perform need not be activities that our minds were designed to perform. We can play chess, program computers, do college-level statistics, and so forth, but these things cannot have been selected for in the Pleistocene: "... the performance on such tasks is generally poor and uneven. In all probability, a wide and somewhat idiosyncratic array of mechanisms and knowledge bases is mobilized when we try to solve this kind of problem, so the study of such problems is unlikely to lead us to carve nature at the joints."

I do not agree that "the performance on such tasks is generally poor and uneven." Most of us can learn to drive a car properly—no mean feat—but we are not primed to do this due to selection in the Pleistocene. Furthermore, Cosmides and Tooby's line of reasoning would also apply to solutions of problems that our Pleistocene ancestors did have to solve. Cooperation in hunting is possible only if we know about the behavior of the game we are hunting, if we are able to anticipate what our co-hunters will do, if we can properly handle our weapons, and so forth. Surely, we need a "wide and somewhat idiosyncratic array of mechanisms and knowledge bases" for this. Indeed, the assumption of evolutionary psychologists that the mind, just like the body, has "organs" appears to represent a fallacious reification (Hendrick 1995).

This points to another problem that evolutionary psychologists have to face. How can we individuate domain-specific problems and mechanisms to solve them? Buss (1995) provides a list of 20 examples. He notes, for example, that phobias of snakes and spiders are common, while fears of weapons and cars are rare. This, according to him, suggests that humans have innate fears of snakes and spiders—or innate tendencies to acquire such fears. But, as Harris and Pashler (1995) note, we might as well postulate that the fears are evoked by any small thing that darts around quickly. Hinde (1995) criticizes other examples mentioned by Buss.

All in all, the speculations of evolutionary psychologists do not represent solid biology. They vastly underestimate how difficult it is to elaborate evolutionary explanations of human behaviors, and they have failed to provide reliable historical reconstructions needed for this.

7.5 CONCLUSIONS

All extant species have originated from processes of evolution, and man is no exception. Yet human beings are different in that they have a sophisticated culture. To some extent, cultural features must be the result of natural selection, but cultural processes also have a life of their own.

Many types of relation exist between natural selection and cultural change. Some would aim to model cultural change as another process of selection and to elaborate an overarching theory of biological and cultural evolution. Such a theory is not feasible, as cultural processes are highly diverse. Cultural processes, rather, call for a natural history approach.

Evolutionary psychologists have focused on the evolutionary origin of our capacity for culture. Their explanations are speculative because it is hard to reconstruct relations between behavioral features of our ancestors and their environments.

8

Against Evolutionary Ethics

8.1 INTRODUCTION

Evolutionary theory has always been a source of inspiration for worldviews and general views of human nature. Considering ethics, some have argued that evolutionary theory is its best foundation. Others cast evolutionary theory in the more negative role of entailing that ethics cannot have a foundation. A more modest view is that evolutionary theory has implications for ethics simply because it sheds light on human nature. All these views originated in the nineteenth century, and they have managed to persist (for a survey, see Farber 1994).

The thesis that evolutionary theory provides the foundation of ethics should come as a surprise. Evolutionary theory is concerned with factual matters, and philosophers have long argued that it is impossible to derive norms from facts. Such derivations amount to a so-called naturalistic fallacy. The term "naturalistic fallacy" refers, in fact, to a thesis due to the philosopher G. E. Moore—that moral features cannot be reduced to natural features—but nowadays it is also used for the is–ought fallacy. For the record, I note that recognition of the is–ought fallacy is almost invariably attributed to the philosopher David Hume, but he did not oppose the derivation of norms from facts (Arnhart 1998: 69–73).

In the next section, I discuss a recent attempt to steer around the naturalistic fallacy. I argue that the attempt fails for the old logical reasons. Attempts to justify ethics by evolutionary considerations are also problematic as they presuppose that evolutionary theory explains morality. The problem is that human beings exhibit both moral and immoral behaviors, depending on their natures and the circumstances. Evolutionary explanations thus would have to specify what processes and situations should foster morality. Evolutionary ethicists have failed to provide the natural history details needed for this.

Subsequent to this, I consider the view that evolutionary theory does not allow of foundations for ethics. I argue that this view must also be rejected. Next, I consider evolutionary explanations of morality in more detail. Finally, I show that evolutionary thinking may be relevant in normative settings that are mostly disregarded in evolutionary ethics.

8.2 DEMISE OF THE NATURALISTIC FALLACY?

Evolutionary biology could be relevant to ethics in several ways. The strongest relevance would be the derivation of ethical principles from evolutionary biology. Most philosophers regard such a derivation as a naturalistic fallacy, which moves from facts to values in a logically improper way.

Some defenders of evolutionary ethics nonetheless aim to develop an evolutionary foundation for ethics. Well known are the attempts of sociobiologists to base ethics on an evolutionary account of social behavior (Lumsden and Wilson 1983; E. O. Wilson 1978). Their view of evolutionary ethics does not include an explanation of how formally to infer values from facts. This makes them an easy target of criticism (Kitcher 1985). Therefore, I disregard the dispute over sociobiology as a possible foundation for ethics. Instead, I focus on the views of authors who have explicitly considered the fact–value problem from a logical point of view.

Some authors argue that features that are deeply embedded in evolved human nature justify morality. Arnhart (1998) has provided the most extensive recent defense of this view. He equates the morally good with the desirable, and the desirable with what contributes to human flourishing. Ethics is natural since it satisfies natural human desires. At least twenty natural desires exist—for example, parental care, sexual mating, friendship, and justice as reciprocity. Here is one of Arnhart's arguments designed to overcome the naturalistic fallacy:

Moore's worry about the "naturalistic fallacy" presumes Kant's separation between factual judgements of what *is* the case and normative judgements of what *ought* to be the case. But this verbal distinction cannot be maintained in moral practice, because every moral judgement presupposes a factual judgment about the satisfaction of human desires as a *reason* for the normative judgment. If "we ought to be just" is an example of a normative judgment, then we could ask, "Why ought we to be just?" If the answer is "because it is right for us to be just," this would still beg the question of why this *is* right for us. Eventually, we must answer that "we ought to be just because justice satisfies some of our deepest desires and thus contributes to our happiness." A Kantian separation between *is* and *ought* would render all normative judgments impotent, because we would have no factual reasons to obey them. [Arnhart 1998: 82–83]

This is an odd argument, because Arnhart's eventual answer is plainly normative. Apart from this, we face the problem that not all desires that appear to be natural fit in with any common morality. Arnhart (1998: 80) states, for example, that courage in war is a human value because human beings have a strong feeling for patriotic loyalty. What, then, about the murder of innocent civilians and rape? These acts are common in wars. So it is reasonable to assume that they result from human desires that many people have under particular circumstances. Arnhart would surely regard them as immoral. Perhaps he would regard them as abnormal. He does exclude abnormal desires from the basis of morality. Morality springs from our moral sense, and those who have no moral sense—psychopaths, for example—live outside the domain of morality. I doubt if this argument would apply to crimes of war. In any case, the distinction between normality and abnormality is difficult to make in many situations.

In commenting on the moral status of psychopaths, Arnhart states that we should not regard Machiavellians as psychopaths:

I agree that people we might identify as Machiavellians, because they employ deception and manipulation for selfish gain, can sometimes attain worldly success. I also agree that because of the success of such people, natural selection might favor some traits of the Machiavellian temperament. But even Machiavellians show self-control and deliberation in the pursuit of their goals—such as wealth, power, and prestige—and therefore they cannot be complete psychopaths. [Arnhart 1998: 223]

The overall flavor of Arnhart's book makes clear that he would not regard deception and manipulation for selfish gain as morally commendable. Yet he grants that evolutionary theory may explain the existence of Machiavellians. Should their behavior count as immoral as they are a minority? That would make morality a matter of statistics, which is a shaky foundation. Mere biology, anyhow, does not help us here.

Arnhart appears to assume that human nature is ultimately benevolent, and that our deepest desires fit in with morality. But common sense indicates that evil streaks are part and parcel of human nature. If evolutionary biology is to explain moral behaviors, it will have to account for immoral behaviors also. We have to specify what processes foster these types of behavior. That calls for a natural history approach that accounts for variability in human behavior. The sweeping claim that morality satisfies our deepest desires, and that these desires are the outcome of selection, will not do.

Richards (1986a, 1989) provides explicit formal arguments to show that reasoning from facts to values is not fallacious. He argues that it is legitimate, for example, to move from the premise "Moral leaders believe abortion is wrong" to the conclusion "Abortion is wrong," provided that we accept a meta-moral inference principle such as "Conclude as sound ethical injunctions what moral leaders preach" (Richards 1986a). Thus, we are entitled to move from an empirical premise to a normative conclusion. A critic may object that the inference principle is essential for the derivation, so that we rely on a normative matter after all. Richards dismisses this objection on the ground that the inference is not a premise. If it were, then we would be left without a principle authorizing the move from premises to conclusion, and the argument would grind to a halt.

However, we may still be asked to justify the inference principle itself. Richards notes that this invites an infinite regress. Considering inference principles, therefore, philosophers ultimately rely on common sense. They resort to intuitively clear cases that represent matters of fact. "Such justifying arguments, then proceed from what people as a matter of fact believe to conclusions about what principles would yield these matters of fact" (Richards 1986a: 284). For example, if moral leaders would state that torturing babies for fun is morally right, we would presumably reject the inference principle.

This method of justification is not confined to ethics. We also need it to justify inference rules of elementary logic—for example, the *modus ponens*. This rule allows us to move from premises with the form "if p, then q" and "p" to the conclusion "q." An example would be the conclusion that Socrates is mortal inferred from the premises that all men are mortal and that Socrates is a man. The inference rule appears to be legitimate since it generates arguments that we would accept as valid. That is, the rule as a matter of fact generates true conclusions if we depart from true premises.

Considering evolutionary biology, Richards asks us to take for granted the premise that natural selection has equipped human beings with dispositions to heed community welfare. He assumes that community members will have formed for themselves simple rules of inference of the sort "From 'action x promotes the community good' conclude 'x ought to be done.'" Such rules allow us to move from empirical principles to normative conclusions. At a higher level, the simple rules can be justified by an argument of this sort: "Since all men evolved to act in accord with the community good . . ., therefore all men ought to act for the community good." This argument, which again moves from factual matters to a norm, is proper in view of an inference rule like "From 'y is enmeshed in causal matrix x' conclude 'y ought to act in x fashion'." This rule explicates the use of the concept of ought.

A heated debate followed the publication of Richards' views (Ball 1988; Cela-Conde 1986; Gewirth 1986, 1993; Hughes 1986; Richards 1986b; Thomas 1986; Trigg 1986; Voorzanger 1987; P. C. Williams 1993). Opponents have charged, for example, that Richards' reasoning is invalid, as natural selection has fostered immoral dispositions in addition to moral ones. I consider this charge in more detail later. First, I argue that, also from a logical point of view, Richards has by no means beaten the naturalistic fallacy.

Let us return to the example of the moral leaders. Richards' reconstruction is as follows. The premise "Moral leaders believe abortion is wrong" allows the conclusion "Abortion is wrong," provided that we accept a meta-moral inference principle such as "Conclude as sound ethical injunctions what moral leaders preach." We can simplify this by considering a particular instantiation of the inference principle: "From the premise 'Moral leaders believe abortion is wrong' infer the conclusion 'Abortion is wrong.'" Richards does not allow us to regard this inference rule as a premise. After all, we need inference rules in addition to premises; without them, our arguments grind to a halt.

Considering this example, I disagree. If we formulate the inference principle as a premise, we get "If moral leaders believe abortion is wrong, then abortion is wrong." If we add the premise "Moral leaders believe abortion is wrong," we can infer "Abortion is wrong." We now have an argument with the form of the *modus ponens* (more accurately, an analog of the *modus ponens* in the normative domain). This reconstruction is more adequate than Richards' reconstruction, because it allows us to distinguish between formal and material matters. The *modus ponens* as an inference rule represents the formal aspect. We need not bother about it since it is a well-entrenched principle of logic. Our attention should focus rather on the new premise, which must be justified. In my view, we can reject it on the basis of other normative claims, but let us grant that appealing to common sense and intuition would be appropriate.

Richards argues that in appealing to common sense and intuition, we resort again to factual matters. Confirming or disconfirming cases concern what people *as a matter of fact* believe, says Richards. He appears to argue as follows. We need to justify the normative claim (inference principle in his terminology) "If moral leaders believe abortion is wrong, then abortion is wrong." Suppose that moral leaders do believe abortion is wrong. Then we can infer that abortion is wrong. We can test this by considering whether people *as a matter of fact* generally believe that abortion is wrong. If this were true (it is not), then our normative claim would be confirmed because it generates an acceptable conclusion. The conclusion, a normative thesis, is again inferred from factual evidence, according to Richards.

This is an elliptical reconstruction. To the hypothesized fact that people believe abortion is wrong I would add the additional, normative premise "If people believe abortion is wrong, then abortion is wrong." This normative premise allows us to infer that abortion is wrong. It plays a covert role in Richards' scheme to confirm the normative claim "If moral leaders believe abortion is wrong, then abortion is wrong." Richards appears to regard this premise as an inference principle. Logicians would reject his reconstruction for lack of a distinction between formal and material matters.

Logic, from the most common point of view, is concerned with formal matters. Hence, to apply logic fruitfully, we should reconstruct arguments such that we can separate out formal and material aspects. Logic furnishes criteria to assess formal aspects; material aspects are covered by evidence in the form of empirical or normative statements. Richards chooses not to use logic in this way. That is his privilege. But it is misleading to suggest that the naturalistic fallacy can be beaten in this way. The term "naturalistic fallacy" is normally used for arguments that are invalid in the sense that no acceptable *formal* rules of inference allow the transition from premises to conclusion. If we broaden the notion of inference rule such that it also covers material aspects of reasoning, then naturalistic fallacies can always be transformed into valid arguments if appropriate rules are added. If we have an invalid argument with *A* as a premise and *B* as a conclusion, we can make it valid by stipulating that "infer *B* from *A*" is a rule of inference. We can also decide to add the premise "If *A*, then *B*." That amounts to the same thing. However, the latter move is more perspicuous, since it allows us to distinguish between formal and material matters.

Richards suggests that his inference rules have the same status as formal rules such as the *modus ponens*. That is not so. Formal rules, unlike his material rules, are justified by formal means. Suppose that we know that two statements with the form "*A* implies *B*" and "*A*" are true. Then the *modus ponens* allows us to infer "*B*." No logician in his right mind would try to confirm the *modus ponens* by checking if *B* is really true. If *B* turned out to be false, we would have to conclude that the premises are false after all.

Contrary to Richards, we should stick to the view that evolutionary theory does not suffice to justify ethics because such a justification would amount to a naturalistic fallacy. To evolutionary theory we have to add normative premises to arrive at normative conclusions. Richards chooses to regard these premises as inference rules. That does not make much difference. We need to justify them anyway. Thus, Richards faces a new problem. If natural selection as a cause of morality justified morality, then it would have to justify immorality as well. It is plausible that fitness-enhancing forms of immoral behavior exist. How, then, could evolutionary biology serve to condemn these forms of behavior?

Several critics of Richards have formulated objections to this effect. Gewirth, for example, notes that natural selection does not suffice to explain morality:

Even if the evolutionary process is a *necessary* condition of moral judgment and action, is it also a *sufficient* condition? Does an appeal to the evolutionary process as such enable us to give a *specific* explanation of human's engaging in morally right conduct *as against* morally wrong conduct? The answer, even on Richards's own account, seems to be negative. For, as he recognizes, what results from the evolutionary process may be morally wrong as well as morally right. . . . But if this is so, then his evolutionary theory of ethics, by appealing to the "evolutionary process," is not *specific* enough to account for its specific subject-matter, which involves the *differentiation* of the morally right from the morally wrong in human action and judgment. [Gewirth 186: 301]

We should take heed here to distinguish explanation from justification. But, on Richards' view, the justification of morality by natural selection does presuppose that selection explains morality. Hence, Gewirth's comment undermines the project of justification.

Here is what Richards has to say about this point.

An inborn commitment to the community welfare, on the one hand, and an aggressive instinct, on the other, are two greatly different traits. In the first, the particular complex of dispositions and attitudes produced by evolution . . . leads an individual to behave in ways that we can generally characterize as acting for the community good; in the second, the behavior cannot be so characterized. Moral "ought"-propositions are not sanctioned by the mere fact of evolutionary formation of human nature, but by the fact of the peculiar formation of human nature we call "moral," which has been accomplished by evolution. ... All meta-level discussions, all attempts to justify ethical frameworks depend on ... inference rules, whose ultimate justification can only be their acceptance by rational and moral creatures. [Richards 1986a: 288–289]

If the evolutionary scenario is true, then man indeed is "ineluctably a moral being." Gewirth thinks this conclusion involves me in a contradiction, since I also admit that men act immorally. [But this is not contradictory:] The claim that man is ineluctably a moral creature means that by virtue of specific evolutionary processes, he has the *capacity* for acting morally. [Richards 1986b: 342]

"Moral 'ought'-propositions are ... sanctioned by ... the fact of the peculiar formation of human nature we call 'moral.'" That sounds like morality being justified by morality. Analogously, we could argue that propositions in praise of murder are sanctioned by the fact of the peculiar formation of human nature we call "immoral." The justification of moral inference rules that appeals to acceptance by moral creatures also appears to have us run around in circles. We need to know about morality in order to identify moral creatures. Appealing to capacities will not help Richards either. Man has the capacity for acting morally, but also the capacity for acting immorally.

Trigg notes that Richards' view, to make sense, must presuppose that moral behavior is an unavoidable outcome of evolution. But that would make the function of morality unclear:

... if human nature has evolved in such a way that we all naturally want to help whoever needs help, the function of morality as an institution seems somewhat obscure. The need for the urgings of morality seem marginal if we are all going to do right anyway because we want to. Even reasoning that we ought to act for the good of the community seems unnecessary if we are predisposed to do so anyway. [Trigg 1986: 332]

Here is Richards' reply:

Trigg ... asks why it is that if we are designed initially for moral behavior, we still have social institutions ... that urge us to virtue. If we are instinctively moral, why "all [the] agonising why we should be moral"? These sorts of objecting queries assume that nature has acted rather unimaginatively in designing man. Trigg forgets that both social animals and human beings have evolved within complex social environments, so that nature, as it were, counts on such environments in realizing her "hoped for" outcomes: fledglings may have the instinct to fly, but it takes an encouraging shove from the mother bird to get them on their way, and nature counts on this. [Richards 1986b: 350]

Indeed, for the fledglings everything works out well. But considering moral behavior, nature is in for a disappointment. The social environment fosters moral behavior and immoral behavior alike. Thus, we are back at the criticism voiced by Gewirth, and we saw that Richards' reply to it is unsatisfactory.

P. C. Williams (1993) and Gewirth (1993), in a similar vein, note that morality as an unavoidable outcome of evolution would be an odd phenomenon, as morality presupposes freedom. Richards (1993: 124–125) grants this point while noting that reconciling determination with freedom is a problem for all ethical theories. Hence, this does not count against his own theory in particular. I disagree. Most ethical theories would allow for freedom of human beings to act morally or immorally. They would resort to normative considerations to infer that freedom ought to be exercised in favor of moral acts. Richards cannot resort to normative considerations. He argues that facts concerning evolution count in favor of moral acts. But, as we saw, this is problematic because evolution allows of both moral and immoral behavior. Normatively inspired theories do not face this particular problem.

Richards' efforts to derive ethics from evolutionary premises are evidently a failure for both logical and substantive reasons. By way of inference rules he assumes that particular actions are justified on the ground that they have been fostered by natural selection. This virtually comes down to charging the evolutionary process itself with moral force. Most researchers would regard this move as counterintuitive because it is at odds with commonly accepted moral values, but exceptions exist in addition to those of Richards.

Rottschaefer and Martinsen (1995) are aware that it amounts to a naturalistic fallacy to infer moral principles from evolutionary theory only. For the inference to work, we need normative premises in addition to empirical premises from evolutionary theory. They propose that we depart from the following normative principle as a premise: "What makes an action good in the most fundamental sense . . ., what justifies it, is that it promotes human adaptations and, thereby, fitness leading normally to S/R [survival and reproduction]" (p. 399). Rottschaefer and Martinsen grant that this principle specifies a necessary condition, but not a sufficient condition for the moral rightness of actions. Biological considerations imply that we need food and clothes, but they do not determine what type of food we should eat or what clothes we should wear. Furthermore, we should recognize the existence of cultural values in addition to the biological value of fitness.

This appears to imply that persons who decide not to have children and to devote all their time to welfare work are not acting morally. If this implication must be rejected, as common sense indicates, the principle does not specify a necessary condition. Furthermore, we must face that, in some situations, immoral actions may enhance fitness also. Fitness-enhancing actions are moral only if they satisfy independent criteria. Hence, fitness drops out as a criterion of morality. Rottschaefer and Martinsen's theory fails to deliver an evolutionary foundation for ethics.

Rottschaefer has subsequently elaborated the theory in a recent book (Rottschaefer 1998). He provides an overview of how different sciences could help to explain capacities involved in moral agency. Evolutionary biology is among the relevant disciplines, as natural selection has been involved presumably in the origin of some capacities. Rottschaefer further argues that evolutionary theory helps to justify morality. He assumes that particular actions are justified in part on the ground that they have been fostered by natural selection. He does not provide a justification for this normative thesis. As I argued, the thesis is problematic because natural selection may also foster immoral actions.

R. Campbell (1996) has come up with a different evolutionary justification of morality. He grants that evolutionary biology cannot justify any particular morality. But it does justify having some morality rather than none.

... the argument rests on the normative but non-moral principle that having some morality rather than none is justified for every member of the group if having some morality rather than none overwhelmingly improves the life prospects of everyone in the group. Since the biological explanation for the existence of morality implies that having some morality rather than none overwhelmingly improves the life prospects of everyone in the group, it follows (given the principle just cited) that having some morality rather than none is justified. [Campbell 1996: 24]

To assess this view, we should know what the phrase "having no morality" means. One possible meaning is "being immoral." But that is indeterminate unless it is contrasted with a particular morality. On common conceptions of morality, incest is wrong, but it is conceivable that some would regard it as morally commendable. Committing incest thus would count as immoral relative to one morality, and as moral relative to a different morality. On this interpretation, then, Campbell's principle is unhelpful. On a different interpretation, having no morality would amount to the absence of explicit principles as guides of behavior. This would mean that people having no morality could, as a matter of fact, avoid incest, refrain from stealing, and so forth. Campbell's premise that morality improves life prospects becomes problematic under this interpretation. All in all, the thesis that evolutionary biology justifies morality in general, without justifying any particular morality, is too indeterminate to be useful.

Evolutionary biology is obviously not a proper tool to defend particular ethical principles. It could, however, be used as a tool to reject principles. Conceivably, it fits this role better. With proper qualifications, most ethicists would endorse the principle "ought implies can." That is, if human beings do not have the capability to act in accordance with a principle, the principle should be rejected. If evolutionary biology entailed that human beings are incapable of exercising totally indiscriminate altruism, we should not endorse an ethics that calls for this kind of altruism. I grant that evolutionary biology could be relevant theoretically in this way. It would be wise, though, to study human capabilities primarily in a more direct way—for example, by psychological or sociological studies.

We saw that Richards and Rottschaefer put too much trust in evolutionary theory as a basis for ethics. J. H. Campbell (1995) goes even further. He presents an extreme theory of evolutionary ethics that envisages a wholesale suppression of moral considerations by evolutionary principles. He argues that evolution, in man, has entered a new phase. The classical mode is adaptive evolution fueled by the survival of the fittest. On top of that we have what he calls generative evolution, which amounts to the survival of successful lineages. Lineages are successful if they have the capability to evolve faster than competitors. We can take evolution in our own hands by genetic engineering and by manipulating the environment, science and technology, and culture. Campbell assumes that groups devoted to the speeding up of their own evolution will come to dominate our species. He is optimistic about their prospects. Thus, he speculates that attempts to increase intelligence may well lead to the equivalent of 25,000 on an IQ scale in ten generations.

Campbell endorses an ethics of generative evolution (EGE), which amounts to the commitment to procreate the lineage that will occupy the leading edge of life from now on. He states that EGE dispenses with the distinction of what should happen and what will happen. Classical ethics assumes that they are different. EGE assumes that what should happen is simply what will happen. EGE is invulnerable to judgment by lesser ethical systems:

If a government legislates against people tampering with their genetic constitution, this will not matter. The generative race merely will be among those groups which will circumvent that impediment. ... Evolution depends upon the fortunes of the singularly successful and not the majority. In this way it differs fundamentally from democratic, statistical, and mechanistic processes with which we are familiar. [J. H. Campbell 1995: 104]

Campbell assumes that generative evolution will initially favor traits that we already hold in high esteem:

I presume that post-human generative evolution will continue to develop the same traits which raised us from beast to human. These are our distinctively "human" characteristics of intellect, communication, emotional commitments to society, culture which directs individuals into activities that produce generative evolution, tools for intervening in evolution and tool-making capability. There is no reason to believe that they are perfected in us. Evolution undoubtedly will also develop additional new generative substrates, but it is not so obvious what these are. [p. 94]

Here we have a process of evolution that is identical to moral progress. Evolutionary biology is not presented as a foundation for a separate domain of ethics. Instead, evolution is infused with morality such that no separate place remains for ethics.

I conjecture that Campbell overrates somewhat our possibilities to alter human nature. That is an empirical issue that I will not elaborate. More important is Campbell's refusal to allow for a separate domain of values. Evolutionary progress is for him the sole founding value of ethics. Yet the second quotation suggests that he cherishes particular values. He assumes that these values will be fostered by generative evolution as a matter of course. But why should that be so? I can imagine that some highly intelligent humans would decide that they are vastly superior to their fellow humans, and that the cause of evolution is best served by simply murdering those less favored with intelligence. Indeed, the sole emphasis on evolutionary progress does not preclude developments like those we witnessed in Nazi Germany. If Campbell would oppose such developments, he had better search for an independent source of morality. Failing that, his approach has implications that most of us would regard as pernicious.

The thesis that evolutionary biology is a proper foundation for ethics has always been suspect because it represents a naturalistic fallacy violating elementary logic. My analysis indicates that recent defenses of the thesis keep flying in the face of elementary logic.

Most of the authors I have discussed assume that natural selection explains morality. Indeed, the program of justification would not make sense without some such assumption. However, the assumption is false. Human beings exhibit moral behaviors and immoral behaviors, depending on their natures and the circumstances. It is reasonable to speculate that both categories of behavior can pass the test of natural selection. To explain any particular form of behavior, evolutionary accounts would have to specify processes and conditions that generate it. That cannot be done without natural history details. Such details are now sorely lacking in evolutionary ethics.

8.3 NO FOUNDATIONS FOR ETHICS?

It is possible, in principle, that evolutionary biology should explain social behavior in animals and in man. Now, social behavior includes moral behavior. This suggests that evolutionary biology may have implications for ethics, albeit not the strong implications that we considered in the previous section. Let us take for granted that altruism allows of an evolutionary explanation of existing forms of altruistic behavior as an adaptive result of natural selection. Let us also grant that the origin of moral rules commending altruism, and other moral rules, can be explained in this way.

Why do we accept these moral rules? If the evolutionary story is right, the short answer appears to be that it is all a matter of natural causes. We are disposed, rationally and emotionally, to accept the rules. The dispositions ultimately result from natural selection. Yet we may feel that this cannot be the whole story. The acceptance of rules also has to do with reasons that justify them. The process of explanation, which unveils causes, should not be confused with the process of justification.

The philosopher Michael Ruse (1995) argues that our belief in the justification of moral rules itself results from natural selection. Particular forms of altruism are adaptive. So it is our belief that rules commending altruism are objectively true. Moral behavior is enhanced by the belief that morality is objective. Selection promotes not only this behavior, but also convictions that strengthen it. No reason exists, Ruse says, to suppose that our normative ethics corresponds with something "out there." Evolution has deluded us collectively into thinking that morality is objective, whereas it is entirely subjective. Hence it cannot be justified.

Ruse knows full well that causal explanations of morality do not suffice to demote justification. Considering knowledge confirmed by regular senses—not least, knowledge produced by science—he does take the opposite stance: "Our eyes are no less an adaptation than is our normative ethics.... They too help in the business of living: for instance, in the avoidance of danger as exemplified by the speeding train heading towards us" (p. 250). Now trains have an objective existence. Trains kill. We know true things about trains, and it is adaptive to know them. We can indeed explain how true knowledge is produced in this case.

In some cases, Ruse argues, a causal analysis of belief production leads to the opposite conclusion. Beliefs may be untrustworthy because they are produced in inappropriate ways. Ruse briefly discusses the following example (pp. 249–250). In the First World War, following the death of loved ones some survivors back home, turning to spiritualism, would receive comforting messages through the ouija board. The departed would tell survivors about their life in a different world, thus comforting the survivors. Ruse argues that we should regard this "knowledge" as nonveridical. He states that the strain of the loss, together with facts about human nature, explains the messages and makes further inquiry superfluous. The beliefs cannot be justified. However understandable, they are false.

Normative ethics is analogous to the ouija board example rather than to the train example, says Ruse. However, he does not show by independent arguments that objective moral truths do not exist. There is a gap in his analysis which is charted well by the general scheme provided by Sober (1994: Chap. 5).

Sober first discusses Hume's thesis that ought-statements cannot be deduced from is-statements. Such a deduction would amount to a naturalistic fallacy. For example, it is improper to infer from the thesis that torturing people for fun causes great suffering that torturing people for fun is wrong. To get at this conclusion, we need the additional premise that it is wrong to cause great suffering. This is a normative premise which, if added to our is-statement, yields a valid deductive argument.

Hume's thesis does not entail subjectivism, the thesis that no ethical statements are true. But we could infer subjectivism from Hume's thesis together with the premise that, if ought-statements cannot be deduced from is-statements, then no ought-statements are true. However, it is not obvious that the premise is reasonable.

Sober next considers an argument for subjectivism which resembles Ruse's ploy. The point of departure is the thesis that we believe particular ethical statements because of our evolution and because of facts about our socialization.

From this it is inferred that no ethical statement is true. But this inference will not do. The argument needs in addition the thesis that the processes that determine what moral beliefs people have are entirely independent of which moral statements (if any) are true. It is unclear why this should be true. We could apply this line of reasoning also to mathematical beliefs. The mathematical beliefs that we have result from our evolution or our socialization. Yet that is not a reason to consider them as purely subjective (see also Kitcher 1985 and Woolcock 1993, who made the same point).

This parallel between ethics and mathematics has been criticized in view of disanalogies (Bradie 1994: 111–112; Murphy 1982: 107–108; Ruse 1993: 155–156; Waller 1996: 252–253). I would reconstruct the situation as follows. In mathematics, and in empirical science, methods exist to justify particular theses. Natural selection may ultimately have caused us to believe that the theses are true, but that is compatible with our having good reasons to accept them as true. The mere fact that selection promotes particular beliefs does not suffice to discredit them. What about beliefs that we cannot justify? Ruse appears to think that such beliefs are discredited by evolutionary explanations. I would say that the primary reason to reject them is that they cannot be justified. Now, we may feel uneasy in rejecting the beliefs if most people accept them. Evolutionary explanations may help us overcome our uneasiness by showing that some beliefs may be false even though they are generally accepted. Natural selection promotes true beliefs if they are adaptive, but also false beliefs if they are adaptive. Evolutionary considerations thus indicate that general acceptance is not a reliable criterion of truth. But it is not a reliable criterion of falsity either.

Ruse suggests that evolutionary biology entails that moral rules are without foundation. In fact, he appears to *assume* that no foundation exists. Evolutionary biology explains, rather, how it would be possible for people to generally accept moral rules if they were without foundation, the point being that accepting the rules is adaptive. But, as I argued, adaptiveness as such does not logically entail either the presence or the absence of foundations.

We need to consider the justification of morals without relying on evolutionary considerations. Justification in moral matters is not inherently more problematic than justification in science. Consider the moral principle that torturing babies for fun is wrong. This principle can be derived from more general claims, about torturing human beings, or about causing suffering, which most people would accept. On the face of it, this is a proper justification. True, the justification invites the question of whether the more general claims can be justified as well. If we justify them by appealing to new principles, we may be asked to justify them also. The process of justification has to end somewhere. Never will we arrive at a justificatory rockbottom that cannot be questioned. The thesis that morality has no foundation does make sense if it is interpreted as the denial that a rockbottom exists. But on this interpretation, science has no foundation either. Nowhere is unlimited justification possible.

Philosophers often search for rockbottoms. The following passage from Hughes (1986) illustrates this:

> ... it is not surprising that sociobiologists have been unimpressed by the doctrine of the naturalistic fallacy, for its implications seem to be absurd. If values cannot be grounded on facts, then presumably they can only be grounded on other values. But these other values can themselves only be grounded on even more fundamental values, and if we are to avoid an infinite regress we end with some fundamental values that cannot be grounded at all. ... Values thus come to be seen as either unreal, or completely arbitrary. ... In either case, ethics as a serious subject for study no longer exists. [Hughes 1986: 306]

Hughes does not realize that his argument should also apply to science. In all domains of knowledge, we have to accept some beliefs without justification.

Waller opposes this analogy between ethics and science with the following argument:

> Science does reach a justificatory limit, just as ethics does. But when science reaches its limit and turns its spade, justification resources are not yet exhausted: one may turn to justification by appeal to basic shared values (a justification within a shared framework of ethical—rather than scientific—principles). For example, a scientist may acknowledge the general principles of science, but still ask: "Is the pursuit of science really (morally) good?" Here the scientist has moved out of science, and the question posed is ethical. It may (or may not) be resolved for the questioner at the ethical level: "Yes, pursuit of science is good, because pursuit of truth is the proper and highest good for humanity." ... But suppose such ethical justifications are challenged. ... In that case there is no further justificatory domain available, and the remaining intractable disputes are enduringly ethical. Such unresolved moral disputes undercut the objectivity of ethics in a way that has no analogy to the queries that arise *within* the given framework of scientific principles. [Waller 1996: 252]

This argument thrives on a fallacy of changing the subject. We may justify a particular thesis of science by inferring it from premises that we assume are true. Next, we may justify the premises, but the process of justification must stop somewhere in science. Waller envisages the further justification outside science that the search for truth is proper. But this is a different issue. The assumption that truth is valuable cannot help us justify the belief that a particular thesis is true. Waller has failed to discredit the analogy of justification between ethics and science.

Ethicists, like other philosophers, have often suggested that justification should be of the rockbottom variety. They have searched for a rockbottom of highly general foundational principles that generate more mundane normative rules. Elsewhere I have argued that the search for generality is misguided (van der Steen and Musschenga 1992). Theories in ethics and science alike should ideally satisfy many methodological criteria in addition to generality. It is impossible for theories to satisfy all criteria at the same time. We have to prioritize criteria. No reason exists why generality should always get the highest priority. In any case, no highly general theory in ethics has ever generated consensus. So we had better be content with normative principles at lower levels of generality which enjoy consensus. Such principles exist, and they have justificatory force.

For ease of exposition I have assumed, in line with the disputes about evolutionary ethics considered here, that justification proceeds in a linear fashion. From this model, we can but end up with unjustified beliefs. Some would argue that the linear model is inappropriate. They would hold that beliefs form networks in which justification can move in all directions. The examples I gave of concrete patterns of justification in ethics can be accommodated not only by the linear view but also by the network-view, which construes justification as coherence among beliefs. Either view should allow for the possibility of proper justifications in ethics.

8.4 THE EXPLANATORY RELEVANCE OF EVOLUTIONARY BIOLOGY

The analyses in the foregoing sections indicate that evolutionary biology does not suffice to justify principles of ethics. Neither does evolutionary biology imply that principles of ethics cannot be justified. Evolutionary thinking could still be relevant for ethics, though, in that it may explain the existence of morality in man. I have already argued in section 8.2 that evolutionary explanations would have to provide natural history details. A general argument to the effect that morality must result from selection because it is adaptive will not do, because immoral behaviors have passed the test of natural selection also. We would have to specify what specific conditions should have favored moral behavior. As yet, no specifications have been forthcoming in evolutionary ethics.

At first sight, the prospects for explanation are bleak, as morality does not sit well with evolution. Natural selection favors egoism in that the behavior of organisms is geared to the maximization of their own fitness. Organisms should not exhibit evolutionarily altruistic behaviors, which decrease their own fitness while increasing the fitness of other organisms. This is seemingly at odds with ethics. It is true that moral behavior should not be equated with altruism, but morality does presuppose that altruism is possible and is often commendable or even obligatory. We are not obliged to be pervasively altruistic, but pervasive selfishness is not compatible with morality.

Many authors (e.g., Rottschaefer 1998; Sober and Wilson 1998) have rightly warned against the identification of evolutionary egoism or altruism with egoism or altruism in the ordinary sense. Evolutionary biology defines these concepts by effects of behaviors on fitness (reproductive success). Egoism and altruism in the ordinary sense primarily concern intentions rather then effects, and their currency is happiness or well-being, for example, not fitness. Hence, ordinary egoism may represent altruism in the evolutionary sense, and ordinary altruism may represent egoism in the evolutionary sense.

Ordinary altruism is compatible with evolutionary biology. But some authors, most notably Sesardic (1995), hold that evolutionary possibilities for ordinary altruism are nonetheless limited, because links exist between ordinary altruism (egoism) and evolutionary altruism (egoism). In chapter 6, I have criticized this view. I have also argued in that chapter that common approaches of altruism in the literature are one-sided. Evolutionary biologists almost always focus on three explanations of evolutionary altruism: group selection, kin selection, reciprocal altruism. Group selection may foster genuine altruism at the level of individuals, but kin selection and reciprocal altruism reduce apparent altruism to egoism. Reciprocal altruism is generally at centre stage in the literature, because the force of group selection is allegedly weak, while kin selection cannot explain altruism toward nonkin. Reciprocal altruism is often put into the context of game theory, where an important issue is to figure out strategies against the possibility of cheating. This imparts a negative flavor on altruism. Here is an example.

The long-term existence of complex patterns of indirect reciprocity may be seen as favoring the evolution of keen abilities, first, to make one's self seem more altruistic than is the case and, second, to influence others to be altruistic in such fashions as to be deleterious to themselves and beneficial to the moralizer, for example, to lead others to invest too much, invest wrongly in the moralizer or his relatives and friends, or invest indiscriminately on a larger scale than would otherwise be the case. [Alexander 1985, quoted from Thompson 1995: 190]

Behaviors that fit this portrayal no doubt exist. But ecologists have also uncovered all sorts of mutualism among animals where cheating is impossible (see chapter 6). Genuine mutualism is often ignored in evolutionary thinking, with the result that selfishness is overrated. On the further, problematic assumption that evolutionary selfishness always carries ordinary selfishness in its wake, we end up with the bleak view that human beings are pervasively selfish. This view is a dogma that keeps cropping up in many disciplines (see chapter 6; see also Wallach and Wallach 1983).

It is conceivable that evolutionary biology is able in principle to explain some aspects of morality. Altruism in the ordinary sense is not evolutionary altruism, but the two varieties of altruism are not entirely unrelated. Morality is not altruism, but it does presuppose the possibility of altruism. Hence, natural selection may have helped to shape our moral behavior.

It is difficult to get beyond this general thesis, which is not very informative. Behavior does not fossilize. Therefore, reconstructions of our behavioral phylogeny are hardly feasible. We simply do not know *which* aspects of morality, if any, have been fostered by natural selection. Some authors have suggested, for example, that selection may well have promoted an increase in intelligence, and that morality is a byproduct of this (Ayala 1987). This would imply that morality, contrary to a common assumption in evolutionary ethics, is not an evolutionary adaptation. Considering many moral principles, we should in any case assume that natural selection is not responsible for their origin. The abolition of slavery is an example.

Considering the explanation of moral and immoral behavior, we need to rely first and foremost on knowledge of human nature in the present. The further program of investigating the evolutionary origin of morality is interesting in its own right, but we must beware thinking that we cannot explain morality without it. Evolutionists at times overrate the importance of evolutionary thinking. Here is an example:

> To go beyond avoiding the oversimplifications and mistakes we already recognize, in discussing underlying mechanisms of behavior . . . is a monumental task in which behaviorists will eventually have to depend heavily on developmental neurophysiologists. In turn, developmental neurophysiologists will have to depend on evolutionary biologists to understand how the various phenomena they analyze can be understood as "mechanisms." Part of the analysis cannot be completed without knowing what each mechanism is evolved to accomplish. . . . Social scientists and philosophers will have to keep up with at least the bare bones of the arguments. [Alexander 1993: 173–174]

I would hope that we need not wait for scenarios from evolutionary biology to understand why people behave in particular ways. Apart from this, I hope that the common emphasis on self-interest, due in part to one-sided evolutionary thinking, will give way to a more balanced view that recognizes genuine mutualism in behavior.

8.5 EVOLUTIONARY THINKING IN NORMATIVE SETTINGS

Evolutionary biology would not suffice as a foundation for ethics. Also, adding to evolutionary principles the normative claim that we should be moral because morality results from natural selection does not generate an acceptable evolutionary ethics. The trouble is that selection may also foster behaviors that we regard as immoral. The normative claim is suspect for this reason.

This does not imply, though, that evolutionary biology has no role to play in normative settings. It deserves a modest but crucial role—for example, in environmental management. Consider the issue of whether the release of genetically modified organisms in the field is hazardous in that it may negatively affect ecosystems or human health (for reference, see Keighery 1995; Marshall 1998; Paoletti and Pimentel 1995; Stephenson and Warnes 1996). Biotechnologists, unlike many ecologists, see few hazards. They have argued, for example, that engineered organisms have a relatively low fitness, so that they will be outcompeted by native species. Disagreements over fitness and competitive ability concern envisaged effects of natural selection, which is an issue in evolutionary ecology. Knowledge of evolutionary processes is essential in the elaboration of normative guidelines in this case.

The use of antibiotics is also a telling example (Poy 1997). Antibiotics have been heralded as wonder drugs ever since their discovery. But their subsequent use on a massive scale is now backfiring. We have been selecting for antibiotic resistance, with the result that the overall effect on human health may well be negative due to new variants of pathogens. Principles of evolutionary ecology would have called for a more modest use of antibiotics.

Ecological thinking is poorly developed in medicine. Garrett (1994) warns that the human species, through factors such as increased population densities and increased migration, is upsetting the ecological balance worldwide. She predicts that infectious diseases will spread as a result, and that the current AIDS epidemic is but the first sign of this. The prediction is based on principles and data of evolutionary ecology which are sorely missing in medicine. Current health-care policies underestimate the hazard of new epidemics.

In all the disputes about relationships between evolutionary biology and ethics, I have never come across links between evolutionary ecology and policies concerning the environment and health care. Yet these links may well represent the most viable and important connection between evolution and ethics. Evolutionary ethicists have been focusing on the wrong issues, unprofitably so.

8.6 CONCLUSIONS

We should not infer values from facts alone. That would amount to a naturalistic fallacy. Some evolutionary ethicists have tried to get around the fallacy. They maintain that evolutionary thinking does provide a foundation for morality. Their attempts to dismiss the naturalistic fallacy are logically flawed. It will not do either to add to evolutionary theory the premise that we should be moral because natural selection has fostered morality. This argument is not a naturalistic fallacy, but it is suspect all the same because selection may foster immoral behaviors also. Evolutionary explanations would have to specify conditions that lead to moral rather than immoral behaviors. Evolutionary ethicists have not provided specifications, which would have to take the form of natural history.

Other evolutionary ethicists have argued that evolutionary biology implies that morality cannot have a foundation. This view presupposes an unacceptable variety of foundationalist thinking. It is obvious that normative principles can be justified, by arguments with other normative principles in the premises. If we aim to justify the premises, we again have to rely on principles, so that we are in for an infinite regress. We must stop the regress somewhere, so that we have to rely on unjustified premises. But if that is taken to imply that justification in ethics is futile, we should bite the further bullet that no domain of knowledge allows of proper justifications. That would be an unpalatable implication. The idea that evolutionary biology militates against a foundation for ethics is therefore implausible.

In principle, evolutionary biology might explain some aspects of morality. In practice, though, such explanations are hard to come by. Explanation is in any case a more modest undertaking than justification.

I would locate the possible relevance of evolutionary thinking for ethics in a different area, which is disregarded in research about evolution and ethics. Policies regarding the environment and health care call for knowledge of evolutionary processes. For example, the assessment of hazards in the release of genetically modified organisms in the environment should account for how natural selection affects modified organisms and native organisms. I suggest that evolutionary ethicists divert their research efforts to issues of this type, which are more important socially than protracted debates about the naturalistic fallacy and speculative evolutionary explanation.

9

Evolution and Knowledge

9.1 INTRODUCTION

Evolutionary biology has been extended to many areas of inquiry. It has invaded the territories of almost all the disciplines concerned with the study of man. Features characterizing man, however special, must have originated in the course of evolution. Evolutionary biologists are therefore entitled to have their say about them. Considering human knowledge, they may be able in principle to explain how it evolved, and how it is used. Epistemology is the discipline primarily concerned with knowledge. At centre stage in this discipline is the question of how we can distinguish between true knowledge and mere belief. That is a normative matter beyond explanation. Evolutionary epistemologists have ventured not only to provide explanations, but also to chart possible implications of evolutionary theory for normative aspects of knowledge. The present chapter reviews evolutionary epistemology, with an emphasis on normative matters.

Bradie (1986, 1989) has usefully distinguished between two programs in evolutionary epistemology; the evolution of cognitive mechanisms program (EEM) and the evolution of theories program (EET). EEM is a straightforward extension of the theory of evolution which focuses on the evolution of traits that play a role in knowledge acquisition. EET attempts to account for the evolution of ideas, scientific knowledge, and culture in general by using models and metaphors drawn from evolutionary biology.

I consider here EEM and EET in this order. Subsequently, I put the problems discussed in evolutionary epistemology in a broader context.

9.2 THE EVOLUTION OF COGNITION

Bradie (1986: 408) rightly notes that "there is a sense in which some version of the EEM program must be true if our current understanding of evolutionary process is anywhere near correct." Features of organisms result from evolutionary processes, and no reason exists to assume that our cognitive abilities are an exception. But this does not tell us much. It does not imply that our cognitive abilities result from evolution *by natural selection*. Neither does it imply that these abilities are adaptive, or that they always promote our getting at the truth.

One way to get at evolutionary explanations of knowledge has been simply to define "knowledge" as a fundamental biological category. D. T. Campbell (1982), for example, extends the "selective paradigm" to all "knowledge processes." He defines these processes as involving three essentials: mechanisms for introducing variation, consistent selection processes, and mechanisms for preserving and/or propagating the selected variations. That is how other people would define evolution. Campbell's terminology does entail that knowledge is the outcome of evolution, because the knowledge process is *defined* as a process of evolution, one that occurs in all organisms, down to unicellular ones. I am afraid that the fundamental problems of epistemology cannot be solved by this kind of definitional move.

I will disregard definitional moves, which have been common in the literature (Riedl 1984 and Heschl 1997 represent other examples). Substantive evolutionary explanations of knowledge will have to do better than this. We need natural history details of the evolutionary processes involved.

Evolutionary epistemology is not limited to explanation. It tends to move from explanation to justification. Kornblith (1985) has provided a nice argument scheme (one that he does not himself endorse) which illustrates how this might be done. Bradie (1989: 403–404), by way of reconstruction, phrases it as follows.

1. Believing truths has survival value.

Therefore,

2. Natural selection guarantees that our innate intellectual endowment gives us a predisposition for believing truths.

Therefore,

3. Knowledge is necessarily a byproduct of natural selection.

4. If nature has so constructed us that our belief-generating processes are inevitably biased in favor of true beliefs, then it must be that the processes by which we arrive at beliefs are those by which we ought to arrive at them.

Therefore,

5. The processes by which we arrive at our beliefs are those by which we ought to arrive at them.

Bradie notes that the argument is not very good. The move from (1) to (2) and (3) is suspect since the fact that a trait has survival value does not ensure that it will be selected for. I would add that (1) is not generally valid. Believing falsehoods may well have survival value under some conditions. Furthermore, Bradie notes, (4) is problematic as the putative selective bias in favor of true beliefs implies nothing about the way in which these beliefs are generated.

Evolutionary epistemology interpreted as a project of justification is alleged to improve on traditional epistemology. Kornblith's argument scheme represents the most plausible reconstruction of the project. It is unsound, substantively and formally (for additional comments on argument schemes of evolutionary epistemology, see Stein 1996: chap. 6). Indeed, to get off the ground, the argument needs the resources of traditional epistemology. Nobody, I guess, would interpret Premise (4) to mean that all our beliefs are true and that our belief-generating processes are always reliable. Hence, epistemology still faces the task of formulating criteria for distinguishing between truths and falsehoods. Would it be possible for evolutionary thinking to generate such criteria? Let us have a try. We could argue that some kinds of true beliefs have survival value, while other kinds of beliefs are less critical for our survival. It is important, for example, for us to know about dangerous snakes, while knowledge about continental drift is less vital. Thus, we could assume that the reliability of belief-generating processes can be judged in relation to the survival value of the beliefs generated (models dealing with this issue are presented in Sober 1994: chap. 3). But this presupposes that we are able to test hypotheses about survival value. So we need to have prior knowledge of how to distinguish between truths and falsehoods.

My comments apply to the entire EEM program. Separate theories differ in details, but they face the same problems. To confirm the general picture, I briefly discuss representative, opposite views of two philosophers, Michael Ruse and Paul Thompson.

Ruse argues that knowledge, as part of human culture, changes rapidly without being tied tightly to adaptive advantage. But it is shaped by underlying principles or norms that belong to our evolutionary heritage. "The norms of knowledge relate to selective advantage" (Ruse 1989: 191). The norms of science, the epitome of knowledge, constitute a generally shared methodology. Scientists, according to Ruse, assume that the world works in a fairly regular sort of way captured by laws of nature which represent causal regularities. Important in the discovery of regularities is the principle of "consilience of inductions" as a criterion of truth. It emphasizes the trustworthiness of hypotheses that fit in with data from separate areas of science.

Ruse (1986: chap. 5) notes that people cling to all sorts of irrational beliefs concerning religion, politics, alternative medicine, and so forth. All these beliefs have their roots in adaptive advantages. For example, "the general urge to take quack medicines is readily explicable in terms of the desperate human urge to maintain life, almost at any cost" (Ruse 1986: 177). Ruse notes that, on the face of it, his overall case becomes a tautology because almost all beliefs thus become compatible with evolutionary epistemology. But it is not really true that anything goes. Clearly, maladaptive beliefs should not exist or should be rare.

I am not sure that adaptation is as widespread as Ruse suggests here. Moreover, it is unclear how evolutionary theory could help us distinguish between rational and irrational beliefs, since both kinds of belief may be promoted by natural selection. To make the distinction, evolutionary explanations would have to specify *which* selection processes foster rational beliefs in particular situations. That calls for natural history, not sweeping claims as endorsed by Ruse.

Ruse (1989) notes that evolutionary epistemologists mostly endorse a realistic picture of the world. After all, it would be odd to explain knowledge as the outcome of evolutionary processes and, at the same time, to regard these processes as mere fabrications occurring in the minds of human beings. This argument only shows that evolutionary epistemology presupposes realism. It fails to justify realism.

Like Ruse, I would endorse common-sense realism. But, unlike Ruse, I do not assume that common-sense realism is supported by evolutionary theory. Ruse notes that philosophical skepticism, which militates against realism, is the most serious problem for evolutionary epistemology (Ruse 1986: 186). But he maintains that several arguments make it innocuous. First, considering scientific progress, he notes that if the skeptic were right to reject the idea of progress, that should not worry the evolutionary epistemologist:

> We must not assume that the best possible analysis of the foundations of science must show us on the road to the certainty that Descartes thought his good God should provide, and for which philosophers ever since have hankered. We are animals, using our evolutionary acquired powers to delve into questions for which such powers were certainly not intended. For the Darwinian, there is no hot line to total truth. If it turns out that the belief in progress is illusory, this will be philosophically disquieting. It will not be a *reductio* of Darwinism. [Ruse 1986: 188]

Ruse is arguing here that evolutionary biology is compatible with the assumption that progress and truth are unattainable in science. This appears to imply that a sound evolutionary foundation for epistemology is impossible! True, Ruse (pp. 200–202) later argues that we have evidence of progress. That is a matter of common sense and proper methodology. But he does not make clear how evolutionary theory could justify this.

I would regard Ruse's next argument as even more devastating for the evolutionary epistemologist.

> Second, complementing this first point, the Darwinian recognizes that philosophical failure to avoid skepticism is quite irrelevant when it comes to the questions which truly count—getting on with survival and reproduction (or even doing science).... the human mind is such that, even if abstract philosophy leads to skepticism, unreasoned optimism keeps us afloat. As human beings, we all believe in the reality

of causality and of the external world and of the worth of consiliences, whatever philosophy might prove. And that is what counts. [p. 188]

This merely introduces common-sense realism as an article of faith. The original point of Ruse's evolutionary epistemology was to forge a link between adaptation and truth. Now he is arguing that it does not matter if skepticism severs the link, because adaptation is all that counts. That does not leave us with much epistemology.

Next, Ruse argues that the skeptic is merely dealing with what is logically possible, not with what it is reasonable and rational to believe. The belief that science is progressive is indeed reasonable and rational because it is a matter of consilience. This must be so simply because "the proper use of consiliences defines what we mean by 'truth' and 'rationality'" (p. 189). I would hold that the evolutionary epistemologist has to show *why* consilience is a proper tool. Ruse decides to regard this as a matter of definition. But definitional fiat is a poor means of justification.

Thompson (1995) has argued that evolutionary epistemology supports anti-realism rather than realism. He notes that natural selection "cares" about survival, not truth. Here is an example.

Suppose that [a] physician treats a certain disease by feeding the patient certain foods. The theoretical explanation is that the demon responsible for the particular disease hates that particular food. Another explanation might appeal to the relation between micro-organisms and certain chemicals in the food. But what is important from the point of view of survival is the practical connection between eating the food and getting well again; the competing theoretical explanations, and more importantly the methods by which these competing theories were formulated and accepted, are really irrelevant here. But even if one grants that our theoretical abilities play an essential role in the development of our practical abilities, this is still no argument for realism. The fact that our various cognitive capacities have contributed to our survival up to now is no argument that they will continue to do so, and more importantly, it is no guarantee that these abilities tend to produce accurate, rather than one of the many possible adequate, pictures of the world. There is no evolutionary argument which goes from our survival either to the survival value or to the truth of our theories and no argument from the survival value of those theories to their truth. [Thompson 1995: 173–174]

This example illustrates that theories are underdetermined by data. The underdetermination thesis constitutes a general philosophical argument against realism. Evolutionary considerations, in Thompson's view, reinforce the argument, because selection often leads to nonadaptive or maladaptive traits. Hence, our perceptual and cognitive capacities are likely not ideally suited for gaining knowledge of the world.

In fact, Thompson's argument only supports the negative point that evolutionary theory does not entail that all our beliefs are true. This does not imply that no means exist to distinguish between true and false beliefs. We can only conclude that evolutionary theory does not provide the proper means. We should rely instead on methodology. That is where the underdetermination thesis comes in. It seems to imply that methodological tools cannot help us devise a realistic picture of the world.

Considering Thompson's example of underdetermination concerning diet and health, I assume that most scientists would reject the demon explanation while accepting the microorganism explanation. The demon explanation is not supported by independent evidence, while the microorganism explanation may fit in with evidence from a variety of sources (e.g., consilience of inductions). Perhaps methodology does not allow us to distinguish between the microorganism explanation and an alternative that is more sophisticated than the demon explanation. Yet, I am not impressed by this problem either. Underdetermination is a problem only if alternative theories explaining data are incompatible. Let us consider a mundane example. The theory that the earth is flat is incompatible with the theory that the earth is a sphere. No scientist would accept the former theory, which has a great variety of data militating against it. Note that the data we have would also count against any other shape for the earth which markedly deviates from a sphere.

My example indicates that *some* theories are so massively supported by evidence that it would be irrational not to accept them. Common sense further indicates that a realistic interpretation of *some* theories is warranted. This is not to deny that other scientific theories—for example, the underdetermination problem—face real problems, so that realistic interpretations may be off the mark.

9.3 THE EVOLUTION OF SCIENTIFIC THEORIES

Regarding the EET program, Hull (1988a, 1988b, 1988c) has provided the most detailed evolutionary theory of science. His strategy is to present a general analysis that applies to all selection processes.

Hull recognizes two types of entities that play a role in all selection processes: replicators and interactors. A replicator is an entity that passes on its structure largely intact in successive replications. An interactor is an entity that interacts as a cohesive whole with its environment in such a way that its interaction causes replication to be differential. Selection is a process of differential extinction and proliferation of interactors which causes the differential perpetuation of the replicators that produced them. As a result of selection, we get lineages—entities that persist indefinitely through time. Genes are paradigm replicators, and organisms are paradigm interactors in natural selection.

Conceptual change in science, according to Hull, likewise involves replicators and interactors. The replicators are elements of the substantive content of science, beliefs in the goals of science, problems and their solutions, modes of representation, data reports, and so forth. Scientists are the primary interactors.

They aim to get their views accepted. That is, they aim to maximize their conceptual inclusive fitness. Hence, science is competitive but also cooperative. If you wish to have your ideas perpetuated, you need the cooperation of colleagues, primarily close colleagues.

Hull argues that individual scientists are motivated by curiosity and the desire to get credit, not primarily by a disinterested search for the truth. Science as an institution is nonetheless progressive and rational due to social interactions among groups. Scientists in other groups than your own have a vested interest in disproving your ideas. Progressiveness is mainly a matter of intergroup competition.

Hull says that he is not concerned with epistemology, by which he means grand views defended in general philosophy, but he deals with epistemology in the form of methodology. This has caused some confusion among commentators, who criticized him on the ground that he implicitly subscribes to an epistemology (see the peer review following Hull 1988a, which is a target article, and Gatens-Robinson 1993). Hull would grant this. He only avoids the label of epistemology because it often represents ambitious undertakings of general philosophy, which he regards as futile.

Considering methodology, Hull assumes that science is progressive and successful in that it uncovers or approximates immutable laws of nature, and that scientists use methodologies to establish a fit between theories and the external world. They test each other's ideas by confronting them with data. Hull's emphasis, though, is more on social factors within science than on methodology. He presents impressive evidence concerning these factors.

The evidence at best provides a partial explanation of the course taken by science, but Hull's aim is not to provide full-fledged explanations of particular episodes in science. He grants that social factors external to science also influence research, but their influence, according to him, cannot be captured by *general* theses. For example, contrary to the view of some "externalists," ideas of scientists do not correlate with political affiliations. Thus, the role of external social factors in science must be covered by particularized case studies—natural history, in my terminology.

"Internalists" hold that the methodology of science suffices to explain progress. Hull agrees with the internalists that critical tests are responsible for convergences of ideas. But he comments that methodology provides no adequate explanation of scientific progress if the social context within science is disregarded.

Hull's own theory aims at generality. He maintains, rightly so I think, that replicators and interactors are proper units to consider at a general level, that scientists as interactors aim to promote their "conceptual inclusive fitness," and that applied methodology in the form of critical tests is vital for progress in science. His evidence confirms that cooperation and competition in science, fueled by the desire for credit, contribute to progress. For example, if editors would consistently exhibit a profound bias against views that deviate from their own ideas, progress would be in jeopardy. Hull's data show that bias is not that widespread. I would add that Hull's data represent valuable natural history with explanatory value. He offers a slender theory containing a limited number of generalities, supplemented with a huge amount of natural history.

Critics have pointed to disanalogies between biological natural selection and conceptual selection as envisaged by Hull. For example, Hull's replicators do not, as in biological evolution, produce interactors in science (Oldroyd 1990; Tennant 1988). This is connected with the lack of an analog of the genotype–phenotype distinction in his theory. Other authors have proposed different evolutionary views of science that make some such distinction. For example, Tennant (1988) regards axioms of a theory as the analog of genes, consequences deduced from axioms as the analog of phenotypic traits, and theories as the analog of organisms. Kantorovich (1989) defends a similar view. But Hahlweg (1989) sees a parallel between genotypes and theories, and between phenotypes and scientific practice. Hahlweg (1988) also argues that Hull overestimates the importance of competitiveness. Fundamental discoveries have been made by scientists who as loners operated in isolation, far away from the arena of competition. Considering analogies with biological evolution, Hahlweg sees parallels with the so-called founder principle: geographic isolation promotes evolutionary change by speciation.

The analogies considered so far concern processes at the population level that result in phylogenetic change. Many evolutionary epistemologists, most notably Campbell (1974) and Popper (1972), have been concerned with ontogeny also. They argue that individual scientists continually generate and test new hypotheses. Tests determine the selective retention or rejection of hypotheses. They resemble trial-and-error learning during ontogeny. But they also fit in with natural selection at the population level and phylogeny. Both ontogeny and phylogeny represent selection processes in which some items are retained while others are rejected.

Bradie (1986) comments that the two sorts of biological processes are very different and that we should be wary of the assumption that the phylogeny and the ontogeny of knowledge can be covered by a single selection model. Amundson (1989) grants that Darwinian natural selection in phylogeny and trial-and-error learning in ontogeny may both allow of a selective explanation, but he opposes selective explanations of scientific change. He argues that the force of a selective explanation depends on the degree to which the following central conditions are met: richness of variation, randomness in the sense of nondirectedness of variation, and a nonpurposive sorting mechanism. Scientific research does not meet these conditions. Scientists do not generate hypotheses randomly. Tests in scientific research involve insights, goals, and intentions of scientists.

Hull would presumably comment that mutations are not generated randomly either. They are at best random, in the sense that mutations do not anticipate adaptive advantages. Furthermore, Hull has argued that the role of intentionality

Evolution and Knowledge

in science is overrated (Hull 1988c: 468–476). Amundson's criticism may also be off target, as he departs from a different conceptualization of selection.

The trouble is that scientific change can be compared with evolution in many different ways. Each comparison yields analogies and disanalogies. The mere fact that disanalogies exist is no reason to dismiss evolutionary epistemology (Bechtel 1989). Analogical reasoning may be fruitful even though analogies are always accompanied by disanalogies, and it occurs throughout science. But the EET program has not been fruitful so far. It has not generated new ideas about science that we would have missed without the analogies. The EET program, like the EEM program, is entirely programmatic.

Furthermore, it is hard to see how evolutionary thinking could yield a normative methodology. Evolutionary theory is concerned with factual matters, and it has no normative implications. EET, like EEM, cannot do without methodology. If we dispense with normative methodology, we do not get epistemology. If we want to have methodology in, evolutionary theory will not be a proper source.

9.4 TOWARD A BROADER PERSPECTIVE

As my comments in the previous sections indicate, I am not positive about evolutionary epistemology. This is not to deny that selection processes in the past may in part explain our cognitive capacities, and that selection processes also play a role in science. In the present section, I put selection processes in science—the subject of EET—in a broader perspective. Regarding EEM, I can only say that, while evolutionary biology is unhelpful relative to normative epistemology, it may in principle help us in the future to develop a descriptive theory about the evolution of cognitive mechanisms.

Considering EET, we should grant that progress in science involves processes of selection. Researchers concerned with EET deserve credit for putting the theme of selection in science on the agenda.

I would distinguish between two levels of selection in science: the selection of subjects to be investigated and the selection of theories about the subjects investigated. Subject selection, nowadays, is to a large extent determined by external factors (e.g., policies of funding agencies). I will come back to that. First, I focus on theory selection.

Theory selection, from a traditional view in the philosophy of science, is ideally a rational affair guided by methodology. But ideals are seldom fully realized, not least because social factors affect theory selection. According to some sociologists of science, this makes methodological standards of rationality obsolete. I disregard this view and assume that, in principle, a rational methodology can be implemented in science (for criticism of extreme sociological views of science, for example, see Brown 1989).

Different methodologies exist in different areas of science—indeed, within a single area. What methodology should we adopt? In biology, Levins (1966, 1968) argued long ago that this question has no general answer. The purposes we have with models and theories determine which methodological criteria are important. In particular, Levins argued, models cannot normally satisfy the criteria of generality, realism, and precision at the same time. If applications of a model we are developing call for great precision, we have to satisfy either generality or realism. I would add that many more methodological criteria exist (e.g., conceptual clarity, fertility, explanatory power, predictive power, testability). This strengthens Levins' view that scientists face methodological trade-offs, and that the priority attached to particular criteria depends on external purposes. Elsewhere I have argued that, by implication, it is legitimate for ethics to play a role at the core of science (van der Steen 1995).

This calls for a new perspective on disputes about realism in evolutionary epistemology and other areas of philosophy. The disputes usually center on the question of whether scientific theories converge on truth or on approximate truth. But that is not a proper question in view of methodological trade-offs. Realism implying truth or approximation of truth is but one among many methodological criteria. Many theories in science are not meant to convey truth in a strict sense. So we have to restrict the question to a subset of theories, those which are aimed to reveal truth. These theories are a mixed lot. Many sophisticated theories are intended to be realistic without our knowing if they are indeed realistic. Other theories are known to be true. The theory that the earth is roughly a sphere is so well confirmed that it is irrational not to accept it as true.

As I said, the ideal of scientific theorizing proceeding by proper methodological guidance is seldom fully realized. It is possible that false ideas keep being proliferated—replicated, if you wish—for lack of adequate methodological scrutiny. The excellent book by Fairhead and Leach (1996) about alleged deforestation in Kissidougou in Guinea provides a nice example. Over a century, politicians and ecologists have argued that the mosaic landscape of forest patches amidst savanna indicates that agricultural practice of farmers in villages is causing progressive deforestation. In line with vegetation succession theory, the assumption has been that forest represents the natural climax community in the area, and that savannas are the result of degrading agricultural practice. Ecologists have done much field work and theoretical work to explain the deforestation. Methodological criticism of this work exists, but the majority opinion has remained that the deforestation is real. The first point to consider here is whether the phenomenon to be explained is indeed real. Fairhead and Leach provide compelling evidence that the amount of forest has actually increased over the last few decades! No sophisticated ecology is needed to show this. Aerial photographs of the area investigated and other records describing the history of the landscape provide sufficient information. The villagers practice a sustainable agriculture, they care for forestation, and they have sound ecological knowledge used to realize forestation. But the government, and development agencies, now try to enforce different, potentially harmful types of agriculture based on the myth of deforestation.

Evolution and Knowledge

By and large, evidence marshaled in support of the deforestation hypothesis concerns features of the biota and the landscape now existing. Historical processes are inferred from this, but historical data—which point in a different direction—are disregarded. Without exception, the evidence is open to alternative interpretations which have been published. But mainstream ecology is unaware of the alternatives, which are spread over many disciplines, and politicians take their lead from mainstream ecology. International development agencies invariably take mainstream ecology for granted also. In Guinea itself, ethnic prejudices work against the rural population. Moreover, urban managers appointed to oversee new agriculture policies aiming to reverse the alleged land degradation have a vested interest in perpetuating the deforestation myth. Their jobs depend on it.

Fairhead and Leach argue that general ecological theory cannot explain the landscape. Interactions among biotic and physical factors are complex, and they cannot be covered by generalities. Instead, we need a painstaking historical approach of local situations. Succession theory, for example, is clearly misguided. It is not true that vegetation always develops toward a stable climax equilibrium. Climate fluctuations and all sorts of natural accidents may prevent the attainment of one particular climax. Sweeping generalizations about the impact of man on the environment are also problematic. Most ecologists assume that this impact can only be negative, particularly as densities of human populations increase. But this is at variance with evidence concerning Guinea and other African countries.

The factors that explain the perpetuation of the deforestation myth are diverse. A few ecologists in prominent positions have promoted the myth, and their ideas have been uncritically accepted by most members of the ecological community. Politicians and scientists alike have not bothered to look at the most direct evidence—aerial photographs and the like. The scientists have inferred historical processes from current characteristics of the land and its biota; in view of published alternative explanations which they disregarded, their inferences are fallacious. The tendency in ecology has been to depart from problematic generalities such as succession theory, while a natural history approach would have been more adequate. Indigenous knowledge has been depreciated or simply disregarded, even though it is superior in this case to commonly accepted views in ecology. Last but not least, opposition to the deforestation myth has not been welcome with international development agencies, and with those in power in Guinea. The opposition would unmask their policies as devastating. Nobody would like such a thing to happen to them. Furthermore, many of those involved have vested interests in perpetuating the deforestation myth.

Fairhead and Leach do research at two levels, science and metascience. They are concerned to chart and explain forestation and deforestation and, at a higher level, to chart and explain existing views about forestation and deforestation. Concerning both levels, they do not aim at highly general theories and explanations. Their approach is best characterized as natural history. It is indeed a fitting approach, since phenomena at either level are influenced by a great diversity of causal factors.

Evolutionary epistemologists dealing with metatheory would describe all this with generalities concerning selection processes. I doubt if this would add any substance to the excellent analysis of Fairhead and Leach. Let us have a try. Considering analogs of genotypes and phenotypes as envisaged by Tennant (1988) and Kantorovich (1989), we could regard succession theory as a genotype which, together with additional assumptions, generates the deforestation hypothesis as a phenotype. Remarkable in this case is the stability of the genotype and the phenotype over a long period. We are obviously dealing with stabilizing selection. The stabilization is odd, since hard evidence counts against both succession theory and the hypothesis of deforestation. It would be natural to construe the evidence as a selective agent that should not allow of the persistence of the genotype and the phenotype. Alternatively, methodological criteria could be cast into the role of selective agents. Considering methodological criteria, we should also note that the trade-off problem involves selection among them, with criteria external to science as selective agents. The different selection processes we get under these analogies are not working the way they should work. In biological natural selection, this would be an incomprehensible phenomenon. But we are dealing now with science, where norms of selection exist in addition to facts of selection. The normative failure of selection is explained by the attitudes of scientists, who disregard the most direct evidence and reason fallaciously from the present situation to historical processes. The scientists are doing the selecting. They are selective agents as well. What about the factors that influence their attitudes? Should they count as selective agents also? In our search for selective agents in science we are apparently facing an embarrassing plethora of possibilities. We have to make a choice that depends on the fruitfulness of the analogies we get. That is, the evolutionary view should teach us something new about science. Considering the example, I am at a loss to say what we could learn. The selection of theories in science is an important phenomenon which is influenced by a heterogeneous array of factors we should know about, but parallels with biological selection will not help us understand the influence of these factors.

Hull's theory would not help us understand the example. Hull is not concerned with the explanation of particular phenomena. He would grant that the tenacity of the deforestation myth must be explained by many different, idiosyncratic factors. And he would state that replication and interaction played a role in this example, as in all other examples of science in the making. I agree, but I am not sure that, when we have explained the example by uncovering idiosyncratic causes, the explanation would be strengthened if we added Hull's generalities to it.

Furthermore, Hull is concerned to explain progress in science, and the example definitely does not constitute progress. The acceptance of a methodologically sound theory and the acceptance of a methodologically unsound theory call for different explanations. If a theory is properly tested by a scientist and it is confirmed, that is a positive factor in getting it accepted. Methodological soundness is not enough, though. The scientists will also need to convince colleagues, and that calls for a proper social setting. An unsound theory gaining credence is an entirely different phenomenon. To explain it, we need to understand what caused the scientist to do methodologically improper work, or how social factors overruled methodological scrutiny in the responses of colleagues.

No common pattern exists for explanations of particular developments in science. We may prepare a checklist of causal factors, but different factors will be operative in different contexts. In the case of the deforestation myth, the most direct evidence was disregarded. That was possible because even this evidence is comparatively indirect, as the hypothesized phenomenon to be explained would take place on a long time-scale. It is not open to direct observation. Other phenomena, say epidemics, are open to more direct inspection, so that it is hardly possible for scientists to fall into the trap of trying to explain a nonphenomenon. The deforestation myth has been influenced by social factors external to science, since deforestation is an important political issue. Other areas of science are not that easily distorted by external factors. For example, politicians and funding agencies do not have much interest in different options to classify organisms, an issue that has been of paramount importance in systematics. Hull's theory concerning evolution in science is geared to the disciplines of systematics and evolutionary biology. He has charted internal social factors affecting progress in systematics in painstaking and admirable detail. Hull states (1988b) that external social factors have effects, but these are idiosyncratic and variable: "If any regularities are to be found in the development of science, they will not be in the area of socioeconomic causes. Instead, the social theory that I set out concerns the *internal* structure of the social relations among scientists and the groups they form" (p. 247). I would argue that we should consider external factors also, if the idiosyncratic and variable effects they have are pronounced. We should resort then to natural history; I regard Hull's emphasis on generality (i.e., "regularities") as misplaced. In any case, Hull may rightly assume that we need not appeal to external factors in explanations of theory change in systematics and evolutionary biology. This may be true, but elsewhere in science such factors are so influential that it would be improper to disregard them.

The example of the deforestation myth is shocking in that it shows that beliefs flying in the face of evidence may be entertained by the scientific community over long periods of time. This invites the question of whether we can devise strategies against such mishaps. The traditional answer would be that a proper application of methodology should suffice, and that we should aim to enhance methodological awareness in science. Indeed we should, but social factors should receive at least as much attention. Considering methodological awareness, we should note that different sciences have assimilated methodology (i.e., philosophical methodology, not statistical methodology) to different extents. It commonly plays an overt role in psychology. But in most areas of biology and biomedicine, explicit methodology is virtually absent. Systematics, Hull's main example, is an exception in that its researchers have explicitly considered methodology over decades. We cannot be sure that deforestation in Guinea would have been exposed as a myth more easily if ecology had contained more explicit methodology. But it is reasonable to assume that biology should often benefit from enhanced methodological awareness.

It is also desirable that influences of external social factors on scientific research get more attention in science itself. For example, researchers outside medicine have shown convincingly that ties between the pharmaceutical industry and the medical community lead to profound bias in biomedical research (Abraham 1995; Payer 1992). Almost never is this issue addressed in medical journals. A recent exception is an article by Stelfox et al. (1998), who reviewed research on calcium-channel antagonists. They demonstrated that the chances that investigators report positive effects of these drugs are way above average if the pharmaceutical industry involved is funding their research. This does not imply that the investigators are villains. They are probably not aware that their work is biased by outside finance. It is to be hoped that other medical researchers will follow the lead of Stelfox et al., and that this kind of bias will become a regular subject of research in biomedicine.

I have distinguished between two overall varieties of selection in science: the selection of subjects to be investigated and the selection of theories about the subjects investigated. So far, my concern was with theory selection. Let us turn to the selection of subjects to be investigated, a theme that most researchers in evolutionary epistemology have disregarded. Nowadays, external social factors determining funding play a major role in this selection. The current emphasis on molecular biology, especially molecular genetics, in biomedical research represents a prominent example of biased funding. In many areas of medicine, especially biological psychiatry, the search is for genetic factors influencing disease. For example, from a widespread view of etiology, genetic abnormalities affecting brain physiology are the primary cause of major psychiatric disorders. Psychosocial factors are regarded as secondary causes that may influence the expression of the genetic factors involved. In chapter 10, I argue that this view of etiology amounts to a methodologically unacceptable bias, because ecology is disregarded. We are dealing with two dichotomies of etiological factors: internal versus external causes and biological versus psychosocial causes. Ecology falls by the wayside due to an amalgamation of the two dichotomies. Research outside mainstream psychiatry convincingly shows that ecological factors affect psychiatric disorders. Diet is an important factor, for example. Food allergies may cause psychosis. Needless to say, patients suffering from this type of allergy should get a proper diet rather than psychopharmaca. As yet, we have no idea how many patients get the wrong treatment due to the existing negligence of ecology.

Bias in subjects to be investigated often entails bias in theory selection. A biased emphasis on molecular biology in biomedicine leads to a biased emphasis on particular etiological factors in theories of pathology. No theory could comprehensively cover all etiological factors affecting a particular disease. Theories focus on a limited set of factors deemed salient. Salience is not an empirical matter. We may wish to focus, for example, on factors that allow of effective treatments since they are manipulable. This makes salience a normative matter.

The example of etiology puts in a new perspective the problem, which I mentioned in the previous section, that theories are underdetermined by data. The underdetermination problem has been a reason for many philosophers to criticize realism, the view that theories tell us something about the real world. They mostly envisage the possibility that a full explanation of some phenomenon may derive in principle from many theories, so that no warrant exists for accepting a particular theory as true. In biology and biomedicine at least, we face a different underdetermination problem. Explanations are almost always partial in that they focus on a limited set of causal factors. Hence, theories are underdetermined by data in the sense that different theories account for different causal backgrounds of a phenomenon, all valid in their own way. Different theories accounting for the same phenomenon may all be realistic in a limited way. Thus, if we wish to select an appropriate theory from a set of theories, salience rather than realism should be the issue. Appropriateness is then a pragmatic matter, because normative factors external to science should determine our choices.

General philosophical views play a visible role in the selection of etiological factors to be investigated in biomedical research. Thus, the emphasis on molecular biology in medicine fits in with the paradigm of reductionism that privileges low levels of organization in the search of fundamental explanations. Apart from this, idiosyncrasies of history play a role. Common approaches in medicine centering on the body can be traced to periods when ecology did not exist in biology. This has led to traditions that are hard to change. In our universities, students of medicine are trained to be content with a restricted form of biology.

A few authors, in excellent recent books, have argued that an urgent need exists for medicine to be infused with ecological thinking (Chivian et al. 1993; Garrett 1994; Pope and Rall 1995). Let us hope that they get a proper hearing in medicine.

Interdisciplinary links are important in all the examples I considered. Fairhead and Leach's work (1996) was successful not least because they managed to integrate methodology, ecology, anthropology, and agricultural science. It is a rare example of generalist and specialist approaches combined. The examples from medicine reveal bias due in part to a shortage of interdisciplinary links. Medicine needs to reckon more with methodology and ecology, and it should pay more attention to the role of external social factors on research. Unfortunately, the present culture does not foster good interdisciplinary work. Scientists are forced to specialize to survive in the publish-or-perish race. The output of science is increasing exponentially, and specialists cannot oversee even what is happening in their own discipline.

A good science policy would have to stimulate generalist work and de-emphasize specialist work. It would have to divert funds from experimental work to the production of reviews, especially reviews that integrate knowledge from different disciplines. But the current culture does not foster this kind of policy.

I do not know how evolutionary epistemologists would deal with developments in science at the level of disciplines. Some of them have regarded theories as the analog of organisms. Others may compare scientists involved in theory selection with organisms in biological selection. Thus, we may regard the body of knowledge of an entire discipline, or the scientific community belonging to a discipline, as the analog of a species. In either case, a disanalogy looms large. Species do not merge, while disciplines do merge or generate interdisciplines. Developments at the level of disciplines are selective, but this selectivity does not have much in common with selection in biological evolution. This confirms that evolutionary epistemology fails to help us understand progress in science.

9.5 CONCLUSIONS

Cognitive capacities in human beings allow them to gather knowledge in special ways. There can be no doubt that these capacities have emerged in the course of evolution. Evolutionary epistemologists have tried to unravel the role of natural selection in the evolution of cognitive capacities. A common assumption is that the capacities have been fostered by natural selection since they are adaptive in that they enable us to attain true beliefs. But details of the selection processes have not been forthcoming. Without details, the assumption is problematic. Adaptive features of organisms need not always result from selection, and nonadaptive features are common. Hence cognitive capacities may be a byproduct of selection for other features, and the capacities need not be pervasively adaptive. Human reasoning at times proceeds in rational ways, but irrational reasoning is also common. Both rational and irrational ways of thinking have passed the test of evolution. Evolutionary explanations of human reasoning, therefore, would have to provide details concerning circumstances that foster rationality or irrationality, and sweeping claims to the effect that it is all a matter of selection need to be replaced by natural history.

The idea that belief-forming processes may result from evolution entails nothing about the reliability of these processes. Evolutionary hypotheses do not provide tools for the justification of beliefs. Now, justification is the central issue for traditional epistemology. Regarding this issue, evolutionary epistemology is singularly unhelpful.

Evolutionary epistemologists have also been concerned with a different project—the explanation of how theories develop, especially in science. It does make sense to describe progress in science as a process of selection. To what extent does this selection process resemble natural selection? That is a moot point. Analogies between the two selection processes can be construed in many different ways. All the analogies are naturally accompanied by disanalogies. That need not bother us if an analogy is fruitful in that it yields new ideas about science. As yet, no analogy has proved to be fruitful in this way. Evolutionary epistemologists' views of science remain disappointingly programmatic.

Both programs of evolutionary epistemology have grappled with the philosophical problem of realism. Does evolutionary analysis warrant the assumption that scientific theories progress toward truth or approximate truth? No satisfactory answer has been forthcoming. Evolutionary epistemology has to draw on the resources of traditional epistemology in the assessment of realism. For example, an evolutionary defence of realism has to presuppose that the theories of evolutionary epistemology themselves are realistic. Thus, we run around in circles.

A traditional view would be that the methodology of science does ensure that theorizing progresses toward the truth. That view is not accepted any more. Social factors have a major impact on theorizing. This may result in irrational beliefs that should not have passed the test of methodological scrutiny. Present-day science has become highly vulnerable to this kind of distortion. The links between scientific research and the industry have become close. Furthermore, the quantity of published work is increasing exponentially. Nobody can oversee his or her own discipline, with the results that relations among disciplines are problematic. These factors easily cause bias in science.

I argue that science should benefit from more emphasis on generalist work—particularly interdisciplinary work—and a de-emphasis of specialist work. Realistic theories are feasible in principle, and adequate methodologies exist to assess realism. To the extent that external social factors may detract from proper theorizing, we need to do more overt research about this. Relations between medicine and the pharmaceutical industry are an example. The industry is a source of biased research, but its influence is seldom considered in medical research articles.

However realistic theories are, they typically cover limited sets of the factors that are responsible for the phenomena studied. Generalist interdisciplinary work can help us be aware of these limitations. Furthermore, we should address the appropriateness or inappropriateness of particular limitations in relation to goals of research.

10

Diseases in an Evolutionary Perspective

10.1 INTRODUCTION

There can be no doubt that the human mind has been shaped in part by natural selection in the past. But evolutionary explanations of particular mental features are, for lack of data concerning their historical development, hard to come by.

Many authors have recently argued that the breakdown of mental functioning in psychopathology is also illuminated by evolutionary thinking. I criticize their view in the present chapter. Malfunctioning and proper functioning of the human mind are equally difficult to explain in evolutionary terms for lack of data.

Other forms of pathology do more easily yield to evolutionary analysis. For example, epidemiological features of infectious diseases depend on short-term processes of natural selection that continually alter characteristics of pathogens. Rapid evolutionary change is possible in pathogens due to their short generation time.

In the next section, I review recent views on the relevance of evolutionary thinking for medicine and psychiatry. Medicine can indeed profit from evolutionary approaches. Subsequently I argue that applications of evolutionary biology to psychopathology are more problematic, as psychiatry has not been sufficiently aware of ecology. If we wish to take ecology into account, we should follow a natural history approach.

10.2 THE NEED FOR DARWINIAN MEDICINE

Nesse and Williams (1994, 1998) have argued convincingly that evolutionary theory explains many features of disease. They distinguish five categories of explanation. First, some discomforting conditions such as fever are evolved defenses, not design defects. Second, some diseases represent conflicts with other organisms, especially pathogens. Third, some changes in circumstances are so recent that natural selection has not had the time to deal with them (the ready availability of dietary fats is an example). Fourth, traits may have both advantages and disadvantages, so that the body falls victim to trade-offs. Fifth, natural selection is restricted by constraints that result in suboptimal design features.

Many evolved defenses exist. Responses such as coughing, sneezing, fever, nausea, vomiting, and diarrhea may help us get rid of infectious microorganisms. Less obvious examples also exist. Bacteria thrive only if they get adequate amounts of iron. That is why the body often reduces blood iron in an evolved defense against invading pathogens. Physicians not trained in evolutionary thinking may not be aware of this and may wrongly interpret reduced iron as a cause of disease. If they prescribe iron as a remedy, they help the pathogens rather than the patient.

Ewald (1994) usefully distinguishes three categories of symptoms of disease: symptoms may represent mere side-effects, or adaptations that benefit the host (defenses by the host), or adaptations that benefit the pathogen (manipulation of the host). Decisions either to treat or not to treat symptoms should be informed by this distinction. Fever in influenza and chickenpox is a telling example (pp. 20–23). Some children who are treated symptomatically with aspirin become confused and delirious; about one-fourth die within days. This is known as Reye's syndrome. Ewald argues that the syndrome may result from one of aspirin's effects, reduced inflammation. The inflammation response is presumably a host defense that helps us overcome the infection.

Conflicts with other organisms, the second category distinguished by Nesse and Williams, also call for an evolutionary perspective. Our evolved defense mechanisms do not make all pathogens harmless for us. That is because the pathogens evolve much faster than we can due to short generation times. They are able to overcome most of our defenses, including defenses in the form of drugs. Antibiotic resistance, which is becoming a major health-care problem, is an example. Superficially, it would seem that the pathogens should benefit from the development of low virulence as the chances of infecting new hosts are poor if infection causes death. Unfortunately, this is not a general rule. Cholera, for example, spreads more rapidly if water supplies are contaminated by immobilized hosts. By implication, sanitation should help to decrease virulence. This is indeed the case. The example illustrates how evolutionary thinking helps us understand patterns of virulence.

Recent changes in circumstances affecting disease, the third category distinguished by Nesse and Williams, also invite an evolutionary approach. Fat, salt, and sugar, as they were scarce, were precious for our ancestors. We still crave for these elements of diet, but nowadays they are readily available. This leads to overconsumption, with all sorts of negative consequences for our health. For example, excessive consumption of fat is a major risk factor in cardiovascular disease.

Trade-offs, the fourth category, are common in evolution. Some genes code for disadvantageous and advantageous phenotype features at the same time. This helps to explain why natural selection does not eliminate all disadvantageous features. Sickle-cell anemia is an example. Persons who are heterozygous for the sickle-cell allele have an enhanced resistance against malaria. This explains the persistence of the anemia in regions where malaria is common.

Lastly, constraints on selection may preclude the elimination of maladaptive features. Natural selection has not managed to eliminate the appendix, a vestigial organ that upon infection may cause death. The cause of this is simple. A decrease in size of the organ would decrease blood flow, which would increase the probability of infection. This works against the selective elimination of the appendix.

Most authors who have considered psychopathology in an evolutionary framework assume that genetic factors are implicated, and that many varieties of psychopathology are so common that they must be, or have been, beneficial in some respects (Nesse and Williams 1994, 1998).

Stevens and Price (1996) hypothesize that schizoid personalities have played a positive role in group splitting in ancestral human populations. Splitting was necessary when groups reached a critical size at which resources were no longer adequate:

At this point, the issue of leadership becomes crucial for survival, because the leader has to inspire the departing group with its sense of mission and purpose, its need to unite against all odds, its belief that it can win through and find its own "promised land." Such a leader needs the sort of charisma traditionally granted by divine will and maintained through direct communion with the gods. It is when called upon to fulfil this exalted role that the schizoid genotype comes into its own. [Stevens and Price 1996: 143]

For example, disordered language of the schizoid leader would help the group assert its identity by the invention of new linguistic forms (p. 149). Furthermore, if it is to split off, the new group must create a new consensus that is at odds with the tradition. Here the schizoid leader, with his genius for delusional and hallucinatory originality, can become indispensable (p. 149). This explanation belongs to Nesse and Williams' third category: a phenotype that was originally adaptive ceases to be adaptive in a new environment.

D. R. Wilson (1993) likewise defends the view that genes implicated in common psychiatric disorders must have been advantageous for our ancestors:

As roles in the kinship band were emerging as a main selective pressure, hominoid neural-humoral systems gradually evolved archetypic small-group processes. Consequently, regulatory norms of humans are those which attune the individual to the social group. From this viewpoint, psychopathology is essentially the study and classification of dissonance in group tuning mechanisms. . . . Such groups of neuropsychologically complex, socially sensitive and intensely interactive anthropoids were advantaged by genotypic characteristics which express pathologic phenotypes only when removed from their typical ecology. . . . [D. R. Wilson 1993: 215–216]

Schlager (1995) and Gilbert (1998) defend similar views.

Considering anxiety disorders, Marks and Nesse (1994) argue that the disorders arise from dysregulation of normal defensive responses that have been promoted by selection in the past (Nesse and Williams' first category). The commonalty of dysregulation is not surprising, since the erring of defensive mechanisms on the side of caution should be adaptive. Nesse (1998) indeed suggests that some seemingly pathological symptoms of mental disorder may still be adaptive. Thus, depression may have the function of helping us disengage from enterprises that will never succeed.

Psychopathology could also represent a trade-off (Nesse and Williams' fourth category). For example, genes responsible for depression and schizophrenia could enhance creativity in patients or in relatives who did not succumb to the disease (Nesse and Williams 1994, 1998).

These explanations share two assumptions. (i) The genetic factors involved are specific in that they code for mental features associated with psychopathology, their expression admittedly depending on environmental factors. (ii) Some mental features linked with psychopathology are advantageous, or they have been advantageous in the past.

In the rest of this chapter, I argue that these assumptions may well be false. The first assumption is often taken for granted as it fits in with a common but problematic paradigm of biological psychiatry. Genes predisposing for psychopathology are assumed to code for enzymes that affect neurotransmitters and thereby impair brain function. This allegedly results in abnormalities of mental functioning and behavior.

Much of the evidence in favor of this paradigm comes from behavior genetics. So let me turn to this discipline.

10.3 NATURAL SELECTION AND PSYCHOPATHOLOGY

Recent research in behavior genetics indicates that genetic factors play important causal roles in human mental functioning and behavior. Pronounced genetic effects have been reported for aggression (Bock and Goode 1996; Stoff and Cairns 1996) and psychiatric disorders (Hall 1996).

Twin studies are an important source of evidence. They suggest that serious psychiatric disorders have major genetic components. Thus, concordance for schizophrenia in monozygotic (MZ) twins (about 50% according to some authors) is much higher than concordance for schizophrenia in dizygotic (DZ)

Diseases in an Evolutionary Perspective

twins. This is plausibly explained by genetic influences because MZ twins, unlike DZ twins, are genetically identical. The explanation presupposes that the so-called equal-environments assumption is valid. That is, MZ and DZ twins are assumed to experience similar correspondences or differences in relevant environmental factors.

Adoption studies confirm this picture. Twins reared apart reveal similar differences in concordance between MZ and DZ pairs. Furthermore, recent research has aimed to identify the genetic factors involved. However, this research is still in its infancy.

Twin studies and adoption studies are fraught with methodological problems (Loehlin 1992). For example, it is difficult to test the equal-environments assumption. Joseph (1998), in a critical analysis, suggests that the assumption is indeed false. Even the finding that MZ twins reared apart show high concordance rates need not implicate genetic factors. The finding may be explained by a shared prenatal environment (Wahlsten and Gottlieb 1997).

For the sake of argument, I assume here that these problems have been overcome. I argue that, even in this ideal situation, evidence from behavior genetics could not tell us much about individuals. Behavior genetics is primarily concerned with population studies, and inferences from populations to individuals are illegitimate. I illustrate this with a hypothetical example. Even the identification of particular genes implicated in phenotype features does not illuminate the origin of these features. We should also know about mechanisms connecting genes and phenotype. Knowledge of mechanisms should include knowledge of the environment. Considering environmental factors, I argue next that behavior genetics, like many areas of medicine and psychiatry, puts a biased emphasis on the psychosocial environment. We sorely need a more balanced, ecological perspective.

Concerning implications of behavior genetics, the following story (from van der Steen 1995, adapted in response to criticism) should have a sobering effect. One hundred pairs of MZ twins are on a trip in a coastal plain. Half of them are short, the others are tall. For the sake of argument, height is here taken to be genetically determined in a strong sense. The trip is wrecked by a flood. The smallish twins drown, while the tall ones survive as they manage to keep their heads above the water. This would be an example of full concordance within twin pairs. So the evidence in favor of genetic determination should be much stronger here than in the case of, say, schizophrenia. We should conclude that the drowning syndrome is genetically determined. Common sense would have it that the flood caused the drowning. Beware of common sense. Sophisticated science shows that it is flawed. Needless to say, this would be an unpalatable conclusion.

We may add DZ twins to the story. DZ twins would show less concordance in drowning since they may differ in height. Also, we could conceive of different trips with floods having different water levels. That would introduce environmental variance in line with the schizophrenia results. For that matter, the story would work for nontwins as well. Suppose that we know about the genes involved. Particular genes influence height and thereby drowning. Yet it seems absurd to regard these genes as a primary cause of drowning.

The story amounts to a *reductio ad absurdum* argument with the following structure.

1. Drowning is genetically determined *in this population* (genetic variance fully accounts for variance in drowning).
2. (From 1): Genetic factors, not the flood, caused drowning.
3. (From common sense): The flood caused drowning.
4. (From 3): Thesis (2) is false.
5. (From 4): If (1) is true, then (2) does not follow from (1).
6. (By analogy): Twin studies say nothing about causes of schizophrenia (or any other feature of individual organisms). Knowledge of the genes involved likewise says nothing about these causes.

The point of the story is that if we think about causes of, for example, schizophrenia, we are normally concerned with the level of individuals, while behavior genetics is concerned with causes at the level of populations. Behavior genetics tells us that variation in schizophrenia in some populations, to a particular degree, is due to genetic sources of variation in addition to environmental sources of variation. The extent of genetic influence at the level of populations is called *heritability*. We should not confuse heritability with *genetic determination* at the level of individuals. Features of individuals are indeed influenced by genetic factors and by environmental factors. But it would be odd to say that genes determine half of John's schizophrenia while the environment determines the other half.

How, then, should we account for the causation of particular features of individuals? That is a question without a general answer. Particular features are influenced by indefinitely many factors. Pragmatic interests determine what factor is regarded as *the* cause or *a* cause of a feature. This implies that the appropriateness of explanations is context-dependent and that we should develop natural history accounts rather than highly general theories.

Here is an additional example, offered by Block (1995). Some years ago, when only women wore earrings, the heritability of having an earring was high because differences in whether a person had an earring were "due" to a chromosomal difference, XX versus XY. Yet we would say that the cultural environment rather than genes has been causally responsible for women wearing earrings. For that matter, now that earrings are less gender-specific, the heritability of having an earring has no doubt decreased.

Considering aggression, Daly (1996: 190) has made a similar point about evidence from behavior genetics.

Familial concordance in aggressivity, for example, does not necessarily mean that there are heritable differences in the psychological mechanisms of aggression; in principle, everyone might be operating on the same contingent decision rule—if bigger than average be mean, say, and if smaller be conciliatory—with the result that whatever engenders heritability of height produces a "reactive" heritability of meanness, too.

Phenylketonuria (PKU) is another telling example. PKU is a disease in which an aberrant gene causes mental retardation due to a person's inability to metabolize the amino acid phenylanaline. Nowadays we know that symptoms of the disease will not develop in persons with the aberrant gene who receive a diet without phenylanaline. Is PKU genetically determined? Yes and no. The answer depends on how we define the "normal" environment. We can attribute the symptoms either to a genetic factor, or to a component of diet. The important point is that we know about physiological processes and environmental influences that lead from a genetic abnormality to PKU symptoms.

My example of twins drowning in a flood also concerns a situation in which our knowledge of relevant factors is adequate. As in the example of PKU, genetic factors are implicated in this example. But our responses to the two examples will be different. In the PKU case, it is natural to regard the aberrant gene as an important causal factor, and the diet therapy aims to correct direct effects of this gene. But considering the drowning, we would not regard the genes involved as important causal factors, and we would not aim to prevent drowning in short people by a growth-hormone therapy.

Regarding effects of genes and the environment on aggression, and on psychopathology, our knowledge is more fragmentary. Hence it is difficult to decide whether these examples resemble the example of PKU or the example of twins in a flood. As I argued, mere knowledge of the magnitude of genetic effects and environmental effects in populations cannot help us devise responsible strategies for dealing with problems at the level of individuals. Yet a covert aim of much behavior genetics is to devise such strategies.

Some researchers have responded, in personal communications, to my examples with the accusation that I am fighting windmills, as behavior geneticists would agree with my points. However, the literature indicates otherwise. Consider the following quotations from sources in behavior genetics:

DNA does not code for fighting, rape, or robbery. Instead, genes contain the blueprint for proteins and enzymes that shape and underlie neurobiological systems that in turn influence higher order behavioral phenotypes. [Carey 1996: 6].

This statement does not overtly address the level of individuals, but it would hardly be meaningful if it did not concern this level. If genes "influenced" fighting, rape, and robbery via a mechanism like the one in my example of the flood, the expression "shape and underlie" would be highly misleading in this context.

... we can anticipate that major genes contributing to alcoholism, Alzheimer disease, bipolar disorder, schizophrenia, and a host of other ailments will soon be in hand. We can also anticipate that these genetic discoveries will provide important insights into the influence of nongenetic factors on psychiatric disease expression, into the basic pathophysiology of brain disruption, and into novel therapeutic interventions. [Kaufmann, Johnson, and Pardes 1996: 24].

This optimistic view, likewise, makes sense only if it applies to the level of individuals. "Contribute to" should be read as a causal attribution at this level.

Our results [of a twin study] are consistent with the previous literature in suggesting that genetic factors play at least a moderate, and perhaps a quite substantial role in the etiology of AI [affective illness]. [Kendler, Pedersen, Neale, and Mathé 1995: 225].

I assume that "etiology" refers to the level of individuals.

... it would be ludicrous to assert that heredity could be the sole cause in the development of any psychological syndrome, including antisocial behavior ... our findings support the contention that there is an inherited predisposition to antisocial behavior ... [Mason and Frick 1994: 318, 320]

Here, "predisposition" can but refer to individuals.

... genetic factors appear to influence aggressive behavior in adolescence. ... The findings on delinquency contrast with those from studies of adult criminality where genetic factors are influential ... [Thapar and McGuffin 1996: 1116].

It is difficult not to read "influence" and "influential" as applying here to the level of individuals.

Plomin and De Fries (1998: 41) are well aware that inferences from populations to individuals are formally invalid, but they state without argument that the inferences are nonetheless reasonable. "Heritability, then, is a way of explaining what makes people different, not what constitutes a given individual's intelligence. In general, however, if heritability for a trait is high, the influence of genes on the trait in individuals would be strong as well." Indeed, it seems to me that the expression "strong influence of genes in individuals" is virtually meaningless.

We should avoid fallacious inferences from populations to individuals. In the next section, I argue that biased views of environmental factors foster such inferences. My focus is on psychopathology and aggression.

10.4 THE ENVIRONMENT IN AGGRESSION AND PSYCHOPATHOLOGY

Much modern research on psychopathology, and also research on aggression, focuses on genetics and on brain pathology. From a common point of view, genetic deficiencies impair processes in the brain involving neurotransmitters, which leads to impaired mental functioning and aberrant behavior. No researcher would deny that environmental factors also play a role. But research concerned with genetics almost always restricts the environment to the psychosocial environment, implicitly if not explicitly. This amounts to the amalgamation of two dichotomies: biological versus psychosocial factors, and internal versus external factors. This amalgamation also exists in other areas of medicine (for sources, see van der Steen and Thung 1988). It is a symptom of a widespread neglect of ecology in medicine (Chivian et al. 1993; Garrett 1994; Pope and Rall 1995).

Here is a typical example. Stoff and Cairns (1996), in their introduction to a book about aggression and violence edited by them, are at pains to dismiss several myths concerning genetic determination. Let me quote what they have to say about two of the myths:

Myth 1. Violence can be reduced to and explained on the basis of disordered biological processes. The relation between a biological substrate and aggressive behavior can be understood in terms of a one-to-one co-ordinate model, a unidirectional causal influence of biology on behavior.

... ecological risk factors such as poverty and exposure to violence in the home and neighbourhood will be shown to have long-standing traumatic impact on brain structure and function as well as subsequent behavioral adaptation. ... Biological variables influence social behavior and, vice versa, social behavior influences biological variables.

This is a myth within a myth. Biological factors are equated with internal factors, and social factors with external factors. Ecology in the primary sense of the term is disregarded.

Myth 3. Biological factors directly cause aggression with "biology" driving aggression.
Biological factors may exert their effects on aggression via intervening processes which, in turn, influence aggression per se; these intervening processes may include proximal social mechanisms, threshold for aggression, impulsivity, cognition, learning, prosocial behaviors, or other biological events. ...

This amounts to the same myth within a myth.

Biases due to a neglect of ecology have not received much attention. Consider the following hypothetical possibility. Studies in behavior genetics indicate that genetic factors account for 50% of the variance in a psychiatric disorder in our culture. A new government manages to effectuate measures that drastically reduce air pollution, it bans unhealthy foods (e.g., refined sugar), and suchlike. It is conceivable that, once this policy comes into effect, the share of genetic factors in the disorder plummets to 10%. Such a phenomenon may make us value genetic factors in a different way.

If persons with a particular mutation in some gene always develop a psychiatric disorder, whatever the environment they are exposed to, we may rightly regard the mutation as a salient cause of the disorder. This case would be like the PKU example. If only persons with this mutation who live in heavily polluted areas were to develop the disorder, then it would be more natural to regard pollution rather then the mutation as a salient cause. This case would be more like my example of twins in a flood.

At the present time, we simply do not know enough about causal factors implicated in psychopathology and, likewise, causal factors implicated in aggressive behavior to decide which picture is more adequate. But recent research does indicate that ecological factors are important. This research is almost entirely disregarded in behavior genetics. Some examples from the literature are the following.

Many components of diet have a profound influence on mood and behavior (Breakey 1997; Christensen 1997; Heseker, Kubler, Pudel, and Westenhofer 1995), in part via effects on neurotransmission (Wurtman 1983; Young 1993; Zeisel 1986). Low-cholesterol diets appear to foster aggression and suicide in animals and man through effects on metabolism of the neurotransmitter serotonin (Ainiyet and Rybakowski 1996; Kaplan et al. 1994; Wardle 1995). The mechanism involved is complex, as two-way interactions exist between fat intake and the serotonin system (Blundell, Lawton, and Halford 1995). Other components of diet likewise influence aggression (Fachinelli, Ison, and Rodriguez Echandia 1996; Dodman et al. 1996; Moeller et al. 1996). Factors of diet may also be implicated in psychiatric disorders such as autism, schizophrenia, and depression (Abou-Saleh and Coppen 1986; Alpert and Fava 1997; Christensen and Christensen 1988; Hibbeln and Salem 1995; Lucarelli et al. 1995).

Other nonpsychosocial environmental factors potentially inducing psychiatric symptoms include coffee (caffeine: Iancu, Dolberg, and Zohar 1996), drugs (e.g., see Marshall and Douglas 1995; Meador 1998; Patten and Love 1997; Sternbach and State 1997), air pollution (Lundberg, 1996), and pathogens (e.g., see Dietrich, Schedlowski, Bode, Ludwig, and Emrich 1998; Smith 1991). Pathogens may have this effect due to a genetic immune deficiency (Morris 1996) or an acquired immune deficiency (Waltrip et al. 1997). Some factors appear to have prenatal effects, as malnutrition or viral infection during pregnancy may adversely affect mental functioning of offspring (Almeida, Tonkiss, and Galler 1996; Brown et al. 1996; Huttunen, Machon, and Mednick 1994; Kunugi et al. 1995). Prenatal events may indeed explain the commonly observed differences in concordance between MZ and DZ twins. Thus, Davis and Phelps (1995) suggest that the probability of shared infections in monochorionic MZ twins exceeds that in dichorionic MZ twins, in line with differences in concordance for schizophrenia. Influences of infection should likewise result in differences between MZ and DZ twins.

The examples show that ecology must be integrated in behavior genetics. Charting effects of variances in genetic factors and environmental factors at the population level tells us little about the causation of mental features and behaviors in individuals. Nor does the mere identification of genes involved help us much. We need to attend to physiological processes intervening between genes and phenotypes, not merely physiological brain processes. And we need to know about the impact of environmental factors, not merely psychosocial environmental factors.

My assessment of etiological factors in psychopathology suggests that current evolutionary approaches may be inadequate. It is commonly assumed that aberrant genetic factors code for psychopathology, and that several psychopathologies are so common that the mental features involved must be beneficial in some respects, or have been beneficial for our ancestors. These assumptions may well be false. Genetic factors explain part of the variation in psychopathology at the population level, but they need not "code for" psychopathology, and any beneficial effects they have need not be primarily mental.

Consider the possibility that diet is a salient etiological factor. Then we could hypothesize that our preference for fat, sugar, and salt, which may have been adaptive in periods when these substances were scarce, induces psychopathology in some persons. That would be in line with one of Nesse and William's categories, but the hypothesized adaptive advantage would relate to diet, not primarily to mental features.

It is also possible that our current diets are unhealthy simply because modern food processing transforms natural foodstuffs into unnatural ones combined with all sorts of chemical additives. This could be made to fit Nesse and Williams' scheme by the thesis that natural selection did not have enough time to produce adaptation to the new foods. But this view has an artificial ring. Human beings cannot survive atomic blasts, or bullets fired through the heart. It would be odd to say that we need more time to adapt to this kind of thing. However much time we have, we could never adapt. Likewise, perhaps, for grossly unnatural foodstuffs.

We are now creating conditions on the earth that threaten all organisms, ourselves included. Some of these conditions are so unnatural that the thesis that we, or other organisms, cannot adapt to them would be an uninformative platitude. This is not to deny that evolutionary approaches offer new insights regarding disease. Nesse and Williams' five categories of evolutionary explanations make sense, and many of their examples are convincing.

The example of psychopathology shows that we need detailed information about many aspects of disorders to decide what types of evolutionary explanation are relevant. It illustrates the need for natural history. Evolutionary explanations of psychopathology are now premature, because our knowledge of etiology is limited. We should first and foremost unravel ecological aspects of psychiatric disorders.

10.5 CONCLUSIONS

In medicine, diseases are explained by their dispositions and environmental factors affecting them. Medicine would profit from a broader framework that also accommodates evolutionary considerations. Considering evolutionary theory, we may wonder why we are plagued by so many diseases. How could natural selection have failed to eliminate them? Several mechanisms exist to explain this. Natural selection has resulted in adaptations that help us fight pathogens. But the pathogens evolve also, at much higher rates. That limits the power of our evolved defenses. The evolutionary perspective is important for medicine not least because it has implications for the treatment of symptoms. To the extent that symptoms represent defenses, it may be unwise to treat them at all. Furthermore, we should realize that medical treatments may lead to selection among pathogens, with unfortunate consequences. Thus, the overuse of antibiotics has resulted in pervasive resistance among bacteria to them.

Some diseases are fostered by changed conditions. A preference for fat consumption was adaptive for our ancestors since fat had been scarce. It is not scarce any more. If we follow natural inclinations nowadays, we are prone to suffer from cardiovascular disease due to excessive fat intake. Diseases can also represent compromises. Sickle-cell anemia persists in regions where malaria is common, as the sickle-cell gene enhances resistance to malaria.

Finally, constraints on selection prevent the elimination of some diseases.

Psychiatric disorders have also been put into an evolutionary framework. From the most common point of view, disorders such as depression, schizophrenia, and anxiety disorders have not been eliminated by selection, as their characteristic features represented adaptations in our ancestors. This view presupposes that genetic factors code for the expression in some environments of mental features associated with psychopathology. This presupposition may well be false. True, genetic variation does explain part of the variation in psychopathology at the population level. But this does not imply that the relevant genes are causally salient at the level of individuals.

Ecological factors have not received the attention they deserve in the explanation of psychopathology. We should know about the role of these factors to get at sensible evolutionary explanations. Such explanations should take the form of natural history.

11

Conclusions

Throughout the book I have argued that general concepts of evolutionary biology are elusive in that they function as placeholders for different notions. Hence, we must do conceptual analysis to come to grips with the substance of evolutionary thinking. A proper understanding of concepts entails that highly general theories are impossible in evolutionary biology. We must be content with "natural history"—that is, particularized knowledge at low levels of generality.

Conceptual pitfalls come in different kinds. First, we have to deal with ambiguous terms that represent different concepts. The notions of survival, fitness, and natural selection are examples. I considered survival and fitness in chapter 2. In evolutionary biology, "survival" commonly means reproductive survival. If it is taken to mean this, the catchphrase "The fittest survive," which some regard as the principle of natural selection, generates problems if "fitness" is also construed as (expected) reproductive survival. It becomes a tautology on this interpretation, and this allegedly creates a vacuity at the core of evolutionary biology. But "survival" can also stand for the persistence of types in populations. Evolutionary biologists would be silly if they aimed to explain survival as reproductive success by fitness as reproductive success. Fortunately, they are not that silly. They rather explain survival as persistence by fitness as reproductive success together with other factors. A large literature is devoted to the tautology problem in evolutionary biology. If we take care to distinguish between the two meanings of "survival," we can put this literature into the wastebasket. Fitness, in its turn, is also ambiguous. Apart from reproductive success, it represents engineering fitness (more about this later). The notion of natural selection is also ambiguous (see chapter 5). In one sense, it stands for processes in which environmental selective agents cause evolutionary change. But evolution by natural selection in a different sense can occur in the absence of selective agents. This ambiguity is seldom recognized in evolutionary biology and the philosophy of biology. It is commonly assumed that selection always involves selective agents.

Second, some general notions of evolutionary biology have little substance because they cover a heterogeneous category of concrete items. Natural selection representing selective agents and engineering fitness are examples (see chapter 2). Natural selection is not a cause of evolutionary change like, say, temperature. It is a placeholder for all sorts of causes, and substantive evolutionary explanations call for a specification of these causes. Engineering fitness is likewise a placeholder, for features of organisms that play a role in particular selection processes. The thesis that differences in engineering fitness explain differential reproductive success should, therefore, be interpreted as the existential claim that there are differences between organisms in particular features which help to explain differential reproductive success. These features, like the environmental causes involved, must be specified in full-fledged explanations. Upon specification, we get natural history rather than general theory.

Third, some general notions of evolutionary biology are tricky because they represent many-place predicates. "Green" is an example of a one-place predicate. We get well-formed sentences if we attach such a predicate to one item, as in "This table is green." "Give" is an example of a three-place predicate. We should connect it to three items, as in "I gave him a table." "Adaptation" and "optimality" are examples of terms that should be attached to more than one item. If we construe them as one-place predicates, we get nonsense. Features of organisms do not represent adaptations or maladaptations, period. They can be adaptive relative to other features, against an entire make-up, in particular environments. This suggests that adaptationism, the thesis that all or most features of organisms are an adaptive result of natural selection, is suspect for conceptual rather than empirical reasons (chapter 3). This also applies for the thesis that all or most behaviors of organisms are optimal (chapter 4). Behaviors can only be optimal relative to a set of constraints (and currencies and strategy sets). From an inclusive interpretation of constraint, the thesis of optimality becomes a boring tautology. Because different, less inclusive notions of constraint are feasible, we can but argue that behaviors are typically optimal relative to some constraints, and nonoptimal relative to different constraints. General claims about optimality are impossible. We should, rather, specify in what respects the behaviors we care to investigate are optimal. Thus, we again get natural history rather than general theory.

Conceptual pitfalls also plague controversies over group selection and species selection (chapter 5). Groups and species can play many different causal roles in evolutionary processes. Different investigators link the labels of group selection and species selection with different causal roles. I regard these ambiguities as a matter of terminological convention. Nothing much hinges on the labels we use, as long as we properly distinguish between different causal processes. This is not to say that group selection and species selection are *merely* terminological issues. The thesis that group selection or species selection in a particular sense play salient roles in evolution is an empirical matter. Controversies over groups and species versus individuals as units of selection have empirical substance. Another dispute over units of selection opposes genes and individual organisms. This dispute has no substance. Both genes and individuals play causal roles in natural selection. Genic selectionists suggest that genes are more important causally. They argue that selection processes are best represented as changes in gene frequencies, as in equations of population genetics. But the equations do not, as such, tell us anything about the real world. Once we interpret them by providing a proper context, reference to causal processes involving individuals becomes unavoidable. The view that individuals rather than genes are causally salient is likewise unacceptable.

Notions of altruism and egoism are ambiguous, and they also represent many-place predicates (chapter 6). This makes meaningless the general thesis that human behaviors are pervasively selfish—psychological egoism—and the general thesis that this is as it should be—ethical egoism. That alone should discredit the view that altruism in the ordinary sense does not sit well with evolutionary biology. We should distinguish between many varieties of altruism and egoism. Extreme forms of altruism that are incompatible with evolutionary biology are rare, and ethicists would not hold that we are morally obliged to go to extremes. Alleged implications of evolutionary biology for ethics are also suspect, because "altruism" and "egoism" have different meanings in biology. In evolutionary biology itself, the emphasis is overmuch on selfishness defined with reference to fitness. Undervalued are many possibilities for animals to perform genuine cooperation.

Human beings are special in that they have culture. This is not to deny that some aspects of culture allow of evolutionary explanations (chapter 7). But adaptationist claims to the effect that all aspects of culture are adaptive outcomes of natural selection must be rejected. Natural selection may cause cultural change, but cultural change as a separate force may also affect patterns of natural selection. Relations between biology and culture come in many different kinds best described in natural history terms. Cultural change has been described as a form of selection analogous to natural selection. We can, indeed, choose to describe it in this way, but the general notion of cultural selection does not explain much as it covers a heterogeneous collection of processes. Again, a natural history approach that attempts to formulate sweeping generalities is more fitting here. Theories aiming to explain progress in science by a general selection theory are problematic for this reason (chapter 9).

Natural selection has no doubt played a role in the genesis of our ethical norms (chapter 8) and our epistemological norms (chapter 9). But it would be difficult to chart this role in any detail. Natural selection in any case cannot be the whole story. For example, the abolition of slavery was the result of cultural forces that had nothing to do with natural selection. Strong forms of evolutionary ethics and evolutionary epistemology have sought to derive our norms from evolutionary theory. That venture is doomed to fail, for the old reason that it amounts to a naturalistic fallacy. Some researchers have argued that evolutionary biology implies that ethical norms, unlike epistemological norms, cannot be justified. No such thing follows from evolutionary biology. In any case, appropriate justifications exist in ethics. We are able validly to derive specific norms from more general normative premises together with empirical premises. True, the process of justification must end somewhere, but that applies to all argumentation, in science, ethics, and any other domain. This does not imply that evolutionary biology has no role to play in normative matters. For example, our ideas about how we should treat infectious diseases should account for natural selection among pathogens. The use of antibiotics on a massive scale flies in the face of evolutionary biology. It is ultimately counterproductive.

The relevance of evolutionary biology for medicine is surveyed in more detail in chapter 10. Evolutionary thinking can help us distinguish, for example, between pathological symptoms of disease calling for treatment and symptoms representing defense reactions of the body. Moderate anemia may represent a defense reaction in infections, because pathogens cannot thrive without iron. Physicians who treat the anemia through iron supplements may not act in the best interests of their patients. In psychiatry, evolutionary thinking has unfortunately taken the form of indefensible adaptationism. There is a widely held assumption that psychiatric disorders such as schizophrenia and depression are so common that in some way they must be adaptive or have been adaptive in the past. It is implied here that specific genetic factors are implicated in the etiology of the disorders. Evidence allegedly supporting this implication is problematic. Twin research, for example, at best implies that genetic variation in part explains variation of the disorders at the population level, but this does not warrant any claim about etiology at the level of individuals.

Conceptual matters loom large in this book. I would grant that empirical matters are ultimately more important than conceptual ones. But without a proper conceptual equipment, we cannot get empirical matters in the right perspective. My approach has been motivated by the conviction that the impact of concepts is undervalued in present-day evolutionary biology.

References

Abou-Saleh, M. T., and A. Coppen (1986). The biology of folate in depression: Implications for nutritional hypotheses of the psychoses. *Journal of Psychiatric Research* 20: 91–101.

Abraham, J. (1995). *Science, Politics and the Pharmaceutical Industry: Controversy and Bias in Drug Regulation.* London: UCL Press.

Ainiyet, J., and J. Rybakowski (1996). Low concentration level of total serum cholesterol as a risk factor for suicidal and aggressive behavior [in Polish]. *Psychiatria Polska* 30: 499–509.

Alexander, R. D. (1985). A biological interpretation of moral systems. *Zygon* 20: 3–20.

Alexander, R. D. (1993). Biological considerations in the analysis of morality. In M. H. Nitecki and D. V. Nitecki (Eds.), *Evolutionary Ethics*. Albany, NY: State University of New York Press, pp. 163–196.

Almeida, S. S., J. Tonkiss, and J. R. Galler (1996). Prenatal protein malnutrition affects the social interactions of juvenile rats. *Physiology and Behavior* 60: 197–201.

Alpert, J. E., and M. Fava (1997). Nutrition and depression: The role of folate. *Nutrition Reviews* 55: 145–149.

Amundson, R. (1989). The trials and tribulations of selectionist explanations. In K. Hahlweg and C. Hooker (Eds.), *Issues in Evolutionary Epistemology*. Albany, NY: State University of New York Press, pp. 413–432.

Amundson, R. (1994). Two concepts of constraint: Adaptationism and the challenge from developmental biology. *Philosophy of Science* 61: 556–578.

Ariza-Ariza, R., M. Mestanza-Peralta, and M. H. Cardiel (1998). Omega-3 fatty acids in rheumatoid arthritis: An overview. *Seminars in Arthritis and Rheumatism* 27: 366–370.

Arnhart, L. (1998). *Darwinian Natural Right: The Biological Ethics of Human Nature*. Albany, NY: State University of New York Press.
Axelrod, R. (1984). *The Evolution of Cooperation*. New York: Basic Books.
Ayala, F. J. (1987). The biological roots of morality. *Biology and Philosophy* 2: 235–252.
Ball, S. (1988). Evolution, explanation, and the fact/value distinction. *Biology and Philosophy* 3: 317–348.
Beatty, J. (1980). Optimal-design models and the strategy of model building in evolutionary biology. *Philosophy of Science* 47: 532–561.
Beatty, J. (1996). Why do biologists argue like they do? *Philosophy of Science* 64: S432–S443.
Beatty, J., and S. Finsen (1989). Rethinking the propensity interpretation: A peek inside Pandora's box. In M. Ruse (Ed.), *What the Philosophy of Biology Is, Essays Dedicated to David Hull*. Dordrecht: Kluwer, pp. 17–30.
Bechtel, W. (1989). An evolutionary perspective on the re-emergence of cell biology. In K. Hahlweg, and C. Hooker (Eds.), *Issues in Evolutionary Epistemology*. Albany, NY: State University of New York Press, pp. 433–457.
Björklund, M. (1996). The importance of evolutionary constraints in ecological time scales. *Evolutionary Ecology* 10: 423–431.
Blackmore, S. (1999). *The Meme Machine*. Oxford: Oxford University Press.
Block, N. (1995). How heritability misleads about race. *Cognition* 56: 99–128.
Blundell, J. E., C. L. Lawton, and J. C. Halford (1995). Serotonin, eating behavior, and fat intake. *Obesity Research* 3: 471S–476S.
Bock, G. R., and J. A. Goode (Eds.) (1996). *Genetics of Criminal and Antisocial Behavior*. Chichester: Wiley.
Boon, S. D., and J. G. Holmes (1991). The dynamics of interpersonal trust: Resolving uncertainty in the face of risk. In R. A. Hinde and J. Groebel (Eds.), *Cooperation and Prosocial Behaviour*. Cambridge: Cambridge University Press, pp. 190–211.
Boyd, R., and P. J. Richerson (1985). *Culture and the Evolutionary Process*. Chicago, IL: University of Chicago Press.
Boyd, R., and P. J. Richerson (1990). Culture and cooperation. In J. J. Mansbridge (Ed.), *Beyond Self-Interest*. Chicago, IL: University of Chicago Press, pp. 113–132.
Bradie, M. (1986). Assessing evolutionary epistemology. *Biology and Philosophy* 1: 401–459.
Bradie, M. (1989). Evolutionary epistemology as naturalized epistemology. In K. Hahlweg and C. Hooker (Eds.), *Issues in Evolutionary Epistemology*. Albany, NY: State University of New York Press, pp. 393–412.
Bradie, M. (1994). *The Secret Chain: Evolution and Ethics*. Albany, NY: State University of New York Press.
Brandon, R. N. (1990). *Adaptation and the Environment*. Princeton, NJ: Princeton University Press.
Brandon, R. N. (1996). Does biology have laws? The experimental evidence. *Philosophy of Science* 64, S444–S457.
Brandon, R. N., J. Antonovics, R. Burian, S. Carson, G. Cooper, P. S. Davies, C. Horvath, B. D. Mishler, R. C. Richardson, K. Smith, K., and P. Thrall

(1994). Discussion: Sober on Brandon on screening-off and the levels of selection. *Philosophy of Science* 61: 475–486.
Breakey, J. (1997). The role of diet and behavior in childhood. *Journal of Paediatrics and Child Health* 33: 190–194.
Brooks, D. R., and D. A. McLennan (1991). *Phylogeny, Ecology, and Behavior: A Research Program in Comparative Biology.* Chicago, IL: Chicago University Press.
Brown, A. S., E. S. Susser, P. D. Butler, A. R. Richardson, C. A. Kaufmann, and J. M. Gorman (1996). Neurobiological plausibility of prenatal nutritional deprivation as a risk factor for schizophrenia. *Journal of Nervous and Mental Diseases* 184: 71–85.
Brown, J. R. (1989). *The Rational and the Social.* London and New York: Routledge.
Burian, R. (1983). Adaptation. In M. Grene (Ed.), *Dimensions of Darwinism.* Cambridge: Cambridge University Press, pp. 287–314.
Burian, R. M., R. C. Richardson, and W. J. van der Steen (1996). Against generality: Meaning in genetics and philosophy. *Studies in History and Philosophy of Science* 27: 1–29.
Buss, D. M. (1991). Evolutionary personality psychology. *Annual Review of Psychology* 42: 459–491.
Buss, D. M. (1995). Evolutionary psychology: A new paradigm for social science. *Psychological Inquiry* 6: 1–30.
Buss, D. M. (1999). *Evolutionary Psychology: The New Science of the Mind*, Boston, MA: Allyn and Bacon.
Byerly, H., and R. Michod (1991). Fitness and evolutionary explanation. *Biology and Philosophy* 6: 1–22.
Campbell, D. T. (1974). Evolutionary epistemology. In P. A. Schilpp (Ed.), *The Philosophy of Karl Popper I.* LaSalle, IL: Open Court, pp. 413–463.
Campbell, D. T. (1982). Evolutionary epistemology. In H. C. Plotkin (Ed.), *Learning, Development, and Culture: Essays in Evolutionary Epistemology.* New York: Wiley.
Campbell, J. H. (1995). The moral imperative of our future evolution. In R. Wesson and P. A. Williams (Eds.), *Evolution and Human Values.* Albany, NY: State University of New York Press, pp. 79–114.
Campbell, R. (1996). Can biology make ethics objective? *Biology and Philosophy* 11: 21–31.
Carey, G. (1996). Family and genetic epidemiology in aggressive and antisocial behavior. In D. M. Stoff and R. B. Cairns (Eds.), *Aggression and Violence: Genetic, Neurobiological and Biosocial Perspectives.* Mahwah, NJ: Erlbaum, pp. 3–21.
Cartwright, N. (1983). *How the Laws of Physics Lie.* Oxford: Oxford University Press.
Cavalli-Sforza, L. L., and M. W. Feldman (1981). *Cultural Transmission and Evolution: A Quantitative Approach.* Princeton, NJ: Princeton University Press.
Cela-Conde, C. J. (1986). The challenge of evolutionary ethics. *Biology and Philosophy* 1: 293–297.

Chivian, E., M. McCally, H. Hu, and A. Haines (1993). *Critical Condition: Human Health and the Environment.* Cambridge, MA: MIT Press.

Christensen, L. (1997). The effect of carbohydrates on affect. *Nutrition* 13: 503–514.

Christensen, O., and E. Christensen (1988). Fat consumption and schizophrenia. *Acta Psychiatrica Scandinavica* 78: 587–591.

Connor, R. C. (1995a). Altruism among non-relatives: Alternatives to the "prisoner's dilemma." *Trends in Ecology and Evolution* 10: 84–86.

Connor, R. C. (1995b). The benefits of mutualism: A conceptual framework. *Biological Reviews* 70: 427–457.

Cosmides, L. (1989). The logic of social exchange: Has natural selection shaped how humans reason? Studies with the Wason selection task. *Cognition* 31: 187–276.

Cosmides, L., and J. Tooby (1992). Cognitive adaptations for social exchange. In J. H. Barkow, L. Cosmides, and J. Tooby (Eds.), *The Adapted Mind: Evolutionary Psychology and the Generation of Culture.* Oxford: Oxford University Press, pp. 163–228.

Cosmides, L., and J. Tooby (1994a). Beyond intuition and instinct blindness: Toward an evolutionary rigorous cognitive science. *Cognition* 50: 41–77.

Cosmides, L., and J. Tooby (1994b). Origins of domain specificity: The evolution of functional organisation. In L. A. Hirschfeld and S. A. Gelman (Eds.), *Mapping the Mind: Domain Specificity in Cognition and Culture.* Cambridge: Cambridge University Press, pp. 85–116.

Cosmides, L., and J. Tooby (1995). From function to structure: The role of evolutionary biology and computational theories in cognitive neuroscience. In M. S. Gazzaniga (Ed.), *The Cognitive Neurosciences.* Cambridge, MA: MIT Press, pp. 1199–1210.

Daly, M. (1996). Evolutionary adaptationism: Another biological approach to criminal and antisocial behavior. In G. R. Bock and J. A. Goode (Eds.), *Genetics of Criminal and Antisocial Behavior.* Chichester: Wiley, pp. 183–191.

Damuth, J., and I. L. Heisler (1988). Alternative formulations of multilevel selection. *Biology and Philosophy* 3: 407–430.

Darden, L., and J. A. Cain (1989). Selection type theories. *Philosophy of Science* 56: 106–129.

Davies, P. S. (1996). Discovering the functional mesh: On the methods of evolutionary psychology. *Minds and Machines* 6: 559–585.

Davis, J. O., and J. A. Phelps (1995). Twins with schizophrenia: Genes or germs? *Schizophrenia Bulletin* 21: 13–18.

Dawkins, R. (1976). *The Selfish Gene.* Oxford: Oxford University Press.

Dawkins, R. (1982). *The Extended Phenotype: The Gene as the Unit of Selection.* San Francisco, CA: Freeman.

De Jong, G. (1994). The fitness of fitness concepts and the description of natural selection. *Quarterly Review of Biology* 69: 3–29.

DeLuca, P., D. Rothman, and R. B. Zurier. (1995). Marine and botanical lipids as immunomodulatory and therapeutic agents in the treatment of rheumatoid arthritis. *Rheumatic Diseases Clinics of North America* 21: 759–777.

Dietrich, D. E., M. Schedlowski, L. Bode, H. Ludwig, and H. M. Emrich (1998). A viro-psycho-immunological disease-model of a subtype affective disorder. *Pharmacopsychiatry* 31: 77–82.

Dodman, N. H., I. Reisner, L. Shuster, W. Rand, U. A. Luescher, I. Robinson, and K. A. Houpt (1996). Effect of dietary protein content on behavior in dogs. *Journal of the American Veterinary Medical Association* 208: 376–379.
Dugatkin, L. A. (1997). *Cooperation Among Animals: An Evolutionary Perspective.* Oxford: Oxford University Press.
Dugatkin, L. A. (1999). *Cheating Monkeys and Citizen Bees.* New York: Free Press.
Dugatkin, L. A., and M. Mesterton-Gibbons (1996). Cooperation among unrelated individuals: Reciprocal altruism, by-product mutualism and group selection in fishes. *BioSystems* 37: 19–30.
Durham, W. H. (1991). *Coevolution: Genes, Culture and Human Diversity.* Stanford, CA: Stanford University Press.
Emlen, J. M. (1966). The role of time and energy in food preference. *American Naturalist* 100: 611–617.
Endler, J. A. (1986). *Natural Selection in the Wild.* Princeton, NJ: Princeton University Press.
Erasmus, U. (1995). *Fats That Heal, Fats That Kill.* Burnaby, BC: Alive Books (third ed.).
Ewald, P. W. (1994). *Evolution of Infectious Disease.* Oxford: Oxford University Press.
Fachinelli, C., M. Ison, and E. L. Rodriguez Echandia (1996). Effect of subchronic and chronic exposure to 5-hydroxytryptophan (5-HTP) on the aggressive behavior induced by food competition in undernourished dominant and submissive pigeons (*Columba livia*). *Behavioral Brain Research* 75: 113–118.
Fairhead, J., and M. Leach (1996). *Misreading the African Landscape: Society and Ecology in a Forest-Savanna Mosaic.* Cambridge: Cambridge University Press.
Farber, P. L. (1994). *The Temptations of Evolutionary Ethics.* Berkeley, CA: University of California Press.
Fox Keller, E. (1987). Reproduction and the central project of evolutionary theory. *Biology and Philosophy* 2: 383–396.
Frank, R. H. (1988). *Passions within Reason.* New York: Norton.
Frank, R. H. (1990). A theory of moral sentiments. In J. J. Mansbridge (Ed.), *Beyond Self-Interest.* Chicago, IL: University of Chicago Press, pp. 71–96.
Garrett, L. (1994). *The Coming Plague: Newly Emerging Diseases in a World Out of Balance.* New York: Farrar, Straus, and Giroux.
Gatens-Robinson, E. (1993). Why falsification is the wrong paradigm for evolutionary epistemology: An analysis of Hull's selection theory. *Philosophy of Science* 60: 535–557.
Gauthier, D. (1987). Reason to be moral? *Synthese* 72: 5–27.
Gewirth, A. (1986). The problem of specificity in evolutionary ethics. *Biology and Philosophy* 1: 297–305.
Gewirth, A. (1993). How ethical is evolutionary ethics? In M. H. Nitecki and D. V. Nitecki (Eds.), *Evolutionary Ethics.* Albany, NY: State University of New York Press, pp. 241–256.
Gilbert, P. (1988). Evolutionary psychopathology: Why isn't the mind better designed than it is? *British Journal of Medical Psychology* 71: 353–373.
Godfrey-Smith, P., and R. C. Lewontin (1993). The dimensions of selection. *Philosophy of Science* 60: 373–395.

Goodwin, B. C. (1994). *How the Leopard Changed Its Spots*. London: Weidenfeld and Nicholson.
Gould, S. J., and R. C. Lewontin (1979). The spandrels of San Marco and the Panglossian paradigm: A critique of the adaptationist programme. *Proceedings of the Royal Society London* 205: 581–598.
Gould, S. J., and E. S. Vbra (1982). Exaptation—A missing term in the science of form. *Paleobiology* 8: 4–15.
Grantham, T. A. (1995). Hierarchical approaches to macroevolution: Recent work on species selection and the "effect hypothesis." *Annual Review of Ecology and Systematics* 26: 301–321.
Gray, R. D. (1987). Faith and foraging: A critique of the "paradigm argument from design," in A. C. Kamil, J. R. Krebs, and H. R. Pulliam (Eds.), *Foraging Behaviour*. New York: Plenum Press, pp. 69–138.
Griffin, M. R. (1998). Epidemiology of nonsteroidal anti-inflammatory drug-associated gastrointestinal injury. *American Journal of Medicine* 104: 23S–29S, 41S–42S.
Griffiths, P. E. (1996a). Darwinism, process structuralism, and natural kinds. *Philosophy of Science* 63, S1–S9.
Griffiths, P. E. (1996b). The historical turn in the study of adaptation. *British Journal for the Philosophy of Science* 47: 511–532.
Griffiths, P. E., and R. D. Gray (1994). Developmental systems and evolutionary explanation. *Journal of Philosophy* 91: 277–304.
Haccou, P., and W. J. van der Steen (1992). Methodological problems in evolutionary biology: IX. The testability of optimal foraging theory. *Acta Biotheoretica* 40: 285–295.
Hahlweg, K. (1988). Epistemology or not? An inquiry into David Hull's evolutionary account of the social and conceptual development of science. *Biology and Philosophy* 3: 187–192.
Hahlweg, K. (1989). A systems view of evolution and evolutionary epistemology. In K. Hahlweg and C. Hooker (Eds.). *Issues in Evolutionary Epistemology*, Albany, NY: State University of New York Press, pp. 45–78.
Hall, L. L. (Ed.) (1996). *Genetics and Mental Illness: Evolving Issues for Research and Society*. New York: Plenum Press.
Hamilton, W. D. (1964). The genetical theory of social behaviour. *Journal of Theoretical Biology* 7: 1–52.
Harris, C. R., and H. E. Pashler (1995). Evolution and human emotions. *Psychological Inquiry* 6: 44–46.
Hartl, D. L., and A. G. Clark (1989). *Principles of Population Genetics*. Sunderland, MA: Sinauer Associates, second edition.
Hendrick, C. (1995). Evolutionary psychology and models of explanation. *Psychological Inquiry* 6: 47–49.
Heschl, A. (1997). Who's afraid of a non-metaphorical evolutionary epistemology? *Philosophia Naturalis* 34, 107–145.
Heseker, H., W. Kubler, V. Pudel, and J. Westenhofer (1995). Interaction of vitamins with mental performance. *Bibliotheca Nutritio et Dieta* 52: 43–55.
Hibbeln, J. R., and N. Salem Jr. (1995). Dietary polyunsaturated fatty acids and depression: when cholesterol does not satisfy. *American Journal of Clinical Nutrition* 62: 1–9.

Hinde, R. A. (1995). The adaptationist approach has limits. *Psychological Inquiry* 6: 50–53.
Holmes, S. (1990). The secret history of self-interest. In J. J. Mansbridge (Ed.), *Beyond Self-Interest*. Chicago, IL, and London: University of Chicago Press, pp. 267–286.
Hughes, W. (1986). Richards' defense of evolutionary ethics. *Biology and Philosophy* 1: 306–315.
Hull, D. L. (1980). Individuality and selection. *Annual Review of Ecology and Systematics* 11: 311–332.
Hull, D. L. (1988a). A mechanism and its metaphysics: An evolutionary account of the social and conceptual development of science. *Biology and Philosophy* 3: 123–155.
Hull, D. L. (1988b). A period of development: A response. *Biology and Philosophy* 3: 241–263.
Hull, D. L. (1988c). *Science as a Process: An Evolutionary Account of the Social and Conceptual Development of Science*. Chicago, IL, and London: University of Chicago Press.
Huttunen, M. O., R. A. Machon, and S. A. Mednick (1994). Prenatal factors in the pathogenesis of schizophrenia. *British Journal of Psychiatry* (Suppl.) 23: 15–19.
Iancu, I., O. T. Dolberg, and J. Zohar (1996). Is caffeine involved in the pathogenesis of combat-stress reaction? *Military Medicine* 161: 230–232.
James, M. J., and L. G. Cleland (1997). Dietary n-3 fatty acids and therapy for rheumatoid arthritis. *Seminars in Arthritis and Rheumatism* 27: 85–97.
Joseph, J. (1998). The equal environments assumption of the classical twin methods: A critical analysis. *Journal of Mind and Behavior* 19: 325–358.
Kantorovich, A. (1989). A genotype-phenotype model for the growth of theories and the selection cycle in science. In K. Hahlweg and C. Hooker (Eds.), *Issues in Evolutionary Epistemology*. Albany, NY: State University of New York Press, pp. 171–184.
Kaplan, J. R., C. A. Shively, M. B. Fontenot, T. M. Morgan, S. M. Howell, S. B. Manuck, M. F. Muldoon, and J. J. Mann (1994). Demonstration of an association among dietary cholesterol, central serotonergic activity, and social behavior in monkeys. *Psychosomatic Medicine* 56: 479–484.
Kaufmann, C. A., J. E. Johnson, and H. Pardes (1996). Evolution and revolution in psychiatric genetics. In L. L. Hall (Ed.), *Genetics and Mental Illness: Evolving Issues for Research and Society*. New York: Plenum Press, pp. 5–28.
Kendler, K. S., N. L. Pedersen, M. C. Neale, and A. A. Mathé (1995). A pilot Swedish twin study of affective illness including hospital- and population-ascertained subsamples: Results of model fitting. *Behavior Genetics* 25: 217–232.
Kirkham, R. L. (1992). *Theories of Truth: A Critical Introduction*. Cambridge, MA: MIT Press.
Kitcher, P. (1982). *Abusing Science: The Case Against Creationism*. Cambridge, MA: MIT Press.
Kitcher, P. (1985). *Vaulting Ambition*. Cambridge, MA: MIT Press.
Kitcher, P. (1989). Explanatory unification and the causal structure of the world. In

P. Kitcher and W. C. Salmon (Eds.), *Scientific Explanation*. Minneapolis, MN: University of Minnesota Press, pp. 410–505.

Kjeldsen-Kragh, J., O. J. Mellbye, M. Haugen, T. E. Mollnes, H. B. Hammer, M. Sioud, and O. Forre (1995). Changes in laboratory variables in rheumatoid arthritis patients during a trial of fasting and one-year vegetarian diet. *Scandinavian Journal of Rheumatology* 24: 85–93.

Kooijman, S. A. L. M. (1993). *Dynamic Energy Budgets in Biological Systems*. Cambridge: Cambridge University Press.

Kornblith, H. (Ed.) (1985). *Naturalizing Epistemology*, Cambridge, MA: MIT Press.

Kunugi, H., S. Nanko, N. Takei, K. Saito, N. Hayashi, and H. Kazamatsuri (1995). Schizophrenia following in utero exposure to the 1957 influenza epidemics in Japan. *American Journal of Psychiatry* 152: 450–452.

Lawton, J. H. (1999). Are there general laws in ecology? *Oikos* 84: 177–192.

Lemos, N. M. (1984). High-minded egoism and the problem of priggishness. *Mind* 93: 542–558.

Lennox, J. G., and B. E. Wilson (1994). Natural selection and the struggle for existence. *Studies in History and Philosophy of Science* 25: 65–80.

Levins, R. (1966). The strategy of model building in population biology. *American Scientist* 54: 421–431.

Levins, R. (1968). *Evolution in Changing Environments*. Princeton, NJ: Princeton University Press.

Lewontin, R. C. (1974). *The Genetic Basis of Evolutionary Change*. New York and London: Columbia University Press.

Lewontin, R. C. (1987). The shape of optimality. In J. Dupré (Ed.), *The Latest of the Best: Essays on Evolution and Optimality*. Cambridge, MA: MIT Press, pp. 151–159.

Lloyd, E. A. (1988). *The Structure and Confirmation of Evolutionary Theory*. New York: Greenwood Press.

Loehlin, J. C. (1992). *Genes and Environment in Personality Development*. London: Sage.

Looren de Jong, H., and W. J. van der Steen (1998). Biological thinking in evolutionary psychology: Rockbottom or quicksand? *Philosophical Psychology* 11: 183–205.

Lucarelli, S., T. Frediani, A. M. Zingoni, F. Ferruzzi, O. Giardini, F. Quintieri, M. Barbato, P. D'Eufemia, and E. Cardi (1995). Food allergy and infantile autism. *Panminerva Medica* 37: 137–141.

Lumsden, C. J., and E. O. Wilson (1981). *Genes, Mind, and Culture*. Cambridge, MA: Harvard University Press.

Lumsden, C. J., and E. O. Wilson (1983). *Promethean Fire*. Cambridge, MA: Harvard University Press.

Lundberg, A. (1996). Psychiatric aspects of air pollution. *Otolaryngology: Head and Neck Surgery* 114: 227–231.

Marks, I. M., and R. M. Nesse (1994). Fear and fitness: An evolutionary analysis of anxiety disorders. *Ethology and Sociobiology* 15: 247–261.

Marshall G. (1998). Herbicide-tolerant crops—real farmer opportunity or potential environmental problem? *Pesticide Science* 52: 394–402.

Marshall, R. D., and C. J. Douglas (1995). Phenylpropanolamine-induced psychosis. Potential predisposing factors. *General Hospital Psychiatry* 17: 457–458.

Mason, D. A., and P. J. Frick (1994). The heritability of antisocial behavior: A meta-analysis of twin and adoption studies. *Journal of Psychopathology and Behavior Assessment* 16: 301–323.

Maynard Smith, J. (1978). Optimization theory in evolution. *Annual Review of Ecology and Systematics* 9: 31–56.

Mayo, D. G., and N. L. Gilinsky (1987). Models of group selection. *Philosophy of Science* 54: 515–538.

McArthur, R. H., and E. R. Pianka (1966). On optimal use of a patchy environment. *American Naturalist* 100: 603–609.

McCauley, D. E., and M. J. Wade (1980). Group selection: The genetic and demographic basis for the phenotypic differentiation of small populations of *Tribolium castaneum. Evolution* 34: 813–820.

Meador, K. J. (1998). Cognitive side effects of medications. *Neurologic Clinics* 16:141–155.

Mera, S. L. (1994). Diet and disease. *British Journal of Biomedical Science* 51: 189–206.

Michod, E. (1999). *Darwinian Dynamics: Evolutionary Transitions in Fitness and Individuality*. Princeton, NJ: Princeton University Press.

Mills, S., and J. Beatty (1979). The propensity interpretation of fitness. *Philosophy of Science* 46: 263–286.

Mithen, S. (1996). *The Prehistory of the Mind: A Search for the Origins of Art, Religion and Science*. London: Thames and Hudson.

Moeller, F. G., D. M. Dougherty, A. C. Swann, D. Collins, C. M. Davis, and D. R. Cherek (1996). Tryptophan depletion and aggressive responding in healthy males. *Psychopharmacology* 126: 97–103.

Morris, J. A. (1996). Schizophrenia, bacterial toxins and the genetics of redundancy. *Medical Hypotheses* 46: 362–366.

Murphy, J. G. (1982). *Evolution, Morality, and the Meaning of Life*. Totowa: Rowman and Littlefield.

Nagel, T. (1970). *The Possibility of Altruism*. Princeton, NJ: Princeton University Press.

Nesse, R. M. (1998). Emotional disorders in evolutionary perspective. *British Journal of Medical Psychology* 71: 397–415.

Nesse, R. M., and G. C. Williams (1998). Evolution and the origins of disease. *Scientific American* (November): 58–65.

Nesse, R. M., and G. C. Williams (1994). *Why We Get Sick*. New York: Times Books.

Nordmann, A. (1990). Persistent propensities: Portrait of a familiar controversy. *Biology and Philosophy* 5: 379–399.

Oldroyd, D. (1990). David Hull's evolutionary model for the progress and process of science. *Biology and Philosophy* 5: 473–487.

Ollason, J. G. (1991). What is this stuff called fitness. *Biology and Philosophy* 6: 81–92.

Orzack, S. H., and E. Sober (1994). Optimality models and the test of adaptationism. *American Naturalist* 143: 361–380.

Oster, G. F., and E. O. Wilson (1978). *Caste and Ecology in the Social Insects*. Princeton, NJ: Princeton University Press.

Oyama, S. (1985). *The Ontogeny of Information: Developmental Systems and Evolution*. Cambridge: Cambridge University Press.

Paoletti, M. G., and D. Pimentel (1995). The environmental and economic costs of herbicide resistance and host-plant resistance to plant pathogens and insects. *Technological Forecasting and Social Change* 50: 9–23.

Parker, G. A., and J. Maynard Smith (1990). Optimality theory in evolutionary biology. *Nature* 348: 27–33.

Patten, S. B., and E. J. Love (1997). Drug-induced depression. *Psychotherapy and Psychosomatics* 66: 63–73.

Payer, L. (1992). *Disease Mongers: How Doctors, Drug Companies, and Insurers Are Making You Feel Sick*. New York: Wiley.

Pierce, G. J., and J. G. Ollason (1987). Eight reasons why optimal foraging theory is a complete waste of time. *Oikos* 49: 111–118.

Plomin, R., and J. C. DeFries (1998). The genetics of cognitive abilities and disabilities. *Scientific American* (May): 40–47.

Pope, A. M., and D. P. Rall (Eds.) (1995). *Environmental Medicine*. Washington, DC: National Academy Press.

Pope, S. J. (1994). *The Evolution of Altruism and the Ordering of Love*. Washington, DC: Georgetown University Press.

Popper, K. R. (1972). *Objective Knowledge: An Evolutionary Approach*. Oxford: Clarendon Press.

Poy, P. H. (1997). Dissemination of antibiotic resistance—genetic engineering at work among bacteria. *Medicine Sciences* 13: 927–933.

Rescher, N. (1987). Rationality and moral obligation. *Synthese* 72: 29–43.

Resnik, D. B. (1988). Survival of the fittest: Law of evolution or law of probability? *Biology and Philosophy* 3: 349–362.

Richards, R. J. (1986a). A defense of evolutionary ethics. *Biology and Philosophy* 1: 265–293.

Richards, R. J. (1986b). Justification through biological faith: A rejoinder. *Biology and Philosophy* 1: 337–354.

Richards, R. J. (1989). Dutch objections to evolutionary ethics. *Biology and Philosophy* 4: 331–343.

Richards, R. J. (1993). Birth, death, and resurrection of evolutionary ethics. In M. H. Nitecki and D. V. Nitecki (Eds.), *Evolutionary Ethics*. Albany, NY: State University of New York Press, pp. 113–131.

Richardson, R. C. (1996). The prospects for an evolutionary psychology: Human language and human reasoning. *Minds and Machines* 6: 541–557.

Richardson, R. C., and R. M. Burian (1992). A defense of propensity interpretations of fitness. *PSA*, East Lansing, Philosophy of Science Association, vol. 1: 349–362.

Ridley, M. (1996). *The Origins of Virtue*. London: Penguin Books.

Riedl, R. (1984). *Biology and Knowledge: The Evolutionary Basis of Reason*. New York: Wiley.

Riedman, M. L. (1982). The evolution of alloparental care and adoption in mammals and birds. *Quarterly Review of Biology* 57: 405–435.

Rothman D., P. DeLuca, and R. B. Zurier (1995). Botanical lipids: Effects on inflammation, immune responses, and rheumatoid arthritis. *Seminars in Arthritis and Rheumatism* 25: 87–96.

Rottschaefer, W. A. (1998). *The Biology and Psychology of Moral Agency*. Cambridge: Cambridge University Press.

References

Rottschaefer, W. A., and D. Martinsen (1995). Really taking Darwin seriously: An alternative to Michael Ruse's darwinian metaethics. In P. Thompson (Ed.), *Issues in Evolutionary Ethics*. Albany, NY: State University of New York Press, pp. 373–408.

Ruse, M. (1986). *Taking Darwin Seriously*. New York: Blackwell.

Ruse, M. (1989). The view from somewhere: A critical defense of evolutionary epistemology. In K. Hahlweg and C. Hooker (Eds.), *Issues in Evolutionary Epistemology*. Albany, NY: State University of New York Press, pp. 185–228.

Ruse, M. (1993). The new evolutionary ethics. In M. H. Nitecki, and D. V. Nitecki (Eds.), *Evolutionary Ethics*. Albany, NY: State University of New York Press, pp. 133–162.

Ruse, M. (1995). *Evolutionary Naturalism*. London: Routledge.

Schlager, D. (1995). Evolutionary perspectives on paranoid disorder. *The Psychiatric Clinics of North America* 18: 263–279.

Schoener, T. W. (1987). A brief history of optimal foraging theory. In A. C. Kamil, J. R. Krebs, and H. R. Pulliam (Eds.), *Foraging Behaviour*. New York: Plenum Press, pp. 5–67.

Schroeder, D. A., J. F. Dovidio, M. E. Sickiby, L. L. Matthews, and J. L. Allen (1988). Empathic concern and helping behavior: egoism or altruism? *Journal of Experimental Social Psychology* 24: 333–353.

Scott, S. (1988). Motive and justification. *The Journal of Philosophy* 85: 479–499.

Sesardic, N. (1995). Recent work on human altruism and evolution. *Ethics* 106: 128–157.

Shimony, A. (1989a). The non-existence of a principle of natural selection. *Biology and Philosophy* 4: 255–273.

Shimony, A. (1989b). Reply to Sober. *Biology and Philosophy* 4: 281–286.

Sloep, P. B., and W. J. van der Steen (1987). The nature of evolutionary theory: The semantic challenge. *Biology and philosophy* 2: 1–15.

Smith, R. S. (1991). The immune system is a key factor in the etiology of psychosocial disease. *Medical Hypotheses* 34: 49–57.

Sober, E. (1984). *The Nature of Selection: Evolutionary Theory in Philosophical Focus*. Cambridge, MA, and London: MIT Press.

Sober, E. (1987). What is adaptationism? In J. Dupré (Ed.), *The Latest on the Best: Essays on Evolution and Optimality*. Cambridge, MA: MIT Press, pp. 105–118.

Sober, E. (1989). Is the theory of natural selection unprincipled? A reply to Shimony. *Biology and Philosophy* 4: 275–279.

Sober, E. (1991). Evolutionary altruism, psychological egoism, and morality: Disentangling the phenotypes. In M. H. Nitecki, and D. V. Nitecki (Eds.), *Evolutionary Ethics*. Albany, NY: State University of New York Press, pp. 199–216.

Sober, E. (1992). Screening-off and the units of selection. *Philosophy of Science* 59: 142–152.

Sober, E. (1993). *Philosophy of Biology*. Oxford: Oxford University Press.

Sober, E. (1994). *From a Biological Point of View*. Cambridge: Cambridge University Press.

Sober, E. (1995–1996). Evolution and optimality: Feathers, bowling balls, and the thesis of adaptationism. *Philosophic Exchange* 26: 41–57.

Sober, E. (1996). Two outbreaks of lawlessness in recent philosophy of biology. *Philosophy of Science* 64: S458–S467.

Sober, E. (in press). The two faces of fitness. In R. Singh, D. Paul, C. Krimbas, and J. Beatty (Eds.), *Thinking about Evolution: Historical, Philosophical, and Political Perspectives*. Cambridge: Cambridge University Press.

Sober, E., and D. S. Wilson (1994). A critical review of philosophical work on the units of selection problem. *Philosophy of Science* 61: 534–555.

Sober, E., and D. S. Wilson (1998). *Unto Others: The Evolution and Psychology of Unselfish Behavior*. Cambridge, MA: Harvard University Press.

Stanley, S. (1979). *Macroevolution: Pattern and Process*. San Francisco, CA: Freeman.

Stein, E. (1996). *Without Good Reason: The Rationality Debate in Philosophy and Cognitive Science*. Oxford: Clarendon Press.

Stelfox, H. T., G. Chua, K. O'Rourke, and A. S. Detsky (1998). Conflict of interest in the debate over calcium-channel antagonists. *New England Journal of Medicine* 338: 101–106.

Stephens, D. W., and J. R. Krebs (1986). *Foraging Theory*. Princeton, NJ: Princeton University Press.

Stephenson, J. R., and A. Warnes (1996). Release of genetically modified microorganisms into the environment. *Journal of Chemical Technology and Biotechnology* 65: 5–14.

Sterelny, K. (1996). The return of the group. *Philosophy of Science* 63: 562–584.

Sterelny, K., and P. Kitcher (1988). The return of the gene. *Journal of Philosophy* 85: 339–361.

Sternbach, H., and R. State (1997). Antibiotics: neuropsychiatric effects and psychotropic interactions. *Harvard Review of Psychiatry* 5: 214–226.

Stevens, A., and J. Price (1996). *Evolutionary Psychiatry: A New Beginning*. London and New York: Routledge.

Stidd, B. M., and D. L. Wade (1995). Is species selection dependent upon emergent characters? *Biology and Philosophy* 10: 55–76.

Stoff, D. M., and R. B. Cairns (Eds.) (1996). *Aggression and Violence: Genetic, Neurobiological and Biosocial Perspectives*. Mahwah, NJ: Erlbaum.

Taylor, P. W. (1975). *Principles of Ethics: An Introduction*. Encino, CA: Dickenson.

Tennant, N. (1988). Theories, concepts and rationality in an evolutionary account of science. *Biology and Philosophy* 3: 224–231.

Thapar, A., and P. McGuffin (1996). A twin study of antisocial and neurotic symptoms in childhood. *Psychological Medecine* 26: 1111–1118.

Thomas, L. (1986). Biological moralism. *Biology and Philosophy* 1: 316–325.

Thompson, P. (1989). *The Structure of Biological Theories*. Albany, NY: State University of New York Press.

Thompson, P. (1995). Evolutionary epistemology and scientific realism. *Journal of Social and Evolutionary Systems* 18: 165–191.

Tooby, J., and L. Cosmides (1992). The psychological foundations of culture. In J. H. Barkow, L. Cosmides, and J. Tooby (Eds.), *The Adapted Mind*. New York: Oxford University Press, pp. 19–136.

Tooby, J., and L. Cosmides (1995). Mapping the evolved functional organisation of mind and brain. In M. S. Gazzaniga (Ed.), *The Cognitive Neurosciences*. Cambridge, MA: MIT Press, pp. 1185–1197.
Trigg, R. (1986). Evolutionary ethics. *Biology and Philosophy* 1: 325–335.
Trivers, R. (1971). The evolution of reciprocal altruism. *Quarterly Review of Biology* 46: 35–57.
van der Steen, W. J. (1991). Natural selection as natural history. *Biology and Philosophy* 6: 41–44.
van der Steen, W. J. (1993). *A Practical Philosophy for the Life Sciences*. Albany, NY: State University of New York Press.
van der Steen, W. J. (1995). *Facts, Values, and Methodology: A New Approach to Ethics*. Amsterdam and Atlanta: Rodopi.
van der Steen, W. J. (1998). Forging links between philosophy, ethics, and the life sciences: A tale of disciplines and trenches. *History and Philosophy of the Life Sciences* 20: 233–248.
van der Steen, W. J., and H. Kamminga (1991). Laws and natural history in biology. *British Journal for the Philosophy of Science* 42: 445–467.
van der Steen, W. J., and A. W. Musschenga (1992). The issue of generality in ethics. *Journal of Value Inquiry* 26: 511–524.
van der Steen, W. J., and P. J. Thung (1988). *Faces of Medicine: A Philosophical Study*. Kluwer: Dordrecht.
van der Steen, W. J., and B. Voorzanger (1984). Methodological problems in evolutionary biology, III: Selection and levels of organization. *Acta Biotheoretica* 33: 199–213.
Vbra, E. (1989). Levels of selection and sorting with special reference to the species level. *Oxford Surveys in Evolutionary Biology* 6: 11–68.
Voorzanger, B. (1987). No norms and no nature—the moral relevance of evolutionary biology. *Biology and Philosophy* 3: 253–270.
Wade, M. J. (1982). The evolution of interference competition by individual, family, and group selection. *Proceedings of the National Academy of Sciences USA* 79: 3575–3578.
Wade, M. J., and D. E. McCauley (1980). Group selection: The phenotypic and genotypic differentiation of small populations. *Evolution* 34: 799–812.
Wahlsten, D., and D. Gottlieb (1997). The invalid separation of effects of nature and nurture: Lessons from animal experimentation. In R. J. Sternberg and E. Grigorenko (Eds.), *Intelligence, Heredity, and Environment*. Cambridge: Cambridge University Press, pp. 163–192.
Wallach, M. A., and L. Wallach (1983). *Psychology's Sanction of Selfishness: The Error of Egoism in Theory and Therapy*. San Francisco, CA: Freeman.
Waller, B. N. (1996). Moral commitment without objectivity or illusion: Comments on Ruse and Woolcock. *Biology and Philosophy* 11: 245–254.
Waltrip, R. W., R. W. Buchanan, W. T. Carpenter, B. Kirkpatrick, A. Summerfelt, A. Breier, S. A. Rubin, and K. M. Carbone (1997). Borna disease virus antibodies and the deficit syndrome of schizophrenia. *Schizophrenia Research* 23: 253–257.
Wardle, J. (1995). Cholesterol and psychological well-being. *Journal of Psychosomatic Research* 39: 549–562.

Waters, C. K. (1986). Natural selection without survival of the fittest. *Biology and Philosophy* 1: 207–225.

Weber, M. (1996). Fitness made physical: The supervenience of biological concepts revisited. *Philosophy of Science* 63: 411–431.

Weber, M. (1999). The aim and structure of ecological theory. *Philosophy of Science* 66: 71–93.

Williams, P. C. (1993). Can beings whose ethics evolved be ethical beings? In M. H. Nitecki, and D. V. Nitecki (Eds.), *Evolutionary Ethics*. Albany, NY: State University of New York Press, pp. 233–239.

Williams, T. D. (1994). Adoption in a precocial species, the lesser snow goose: Intergenerational conflict, altruism or a mutually beneficial strategy? *Animal Behaviour* 47: 101–107.

Wilson, D. R. (1988). Evolutionary epidemiology and manic depression. *British Journal of Medical Psychology* 71: 375–395.

Wilson, D. R. (1993). Evolutionary epidemiology: Darwinian theory in the service of medicine and psychiatry. *Acta Biotheoretica* 41: 205–218.

Wilson, D. S. (1994). Adaptive genetic variation and human evolutionary psychology. *Ethology and Sociobiology* 15: 219–235.

Wilson, D. S., and E. Sober (1994). Reintroducing group selection to the human behavioral sciences. *Behavioral and Brain Sciences* 17: 585–608.

Wilson, E. O. (1978). *On Human Nature*. Cambridge, MA: Harvard University Press.

Wolfe, M. M. (1996). NSAIDs and the gastrointestinal mucosa. *Hospital Practice* 31: 37–44, 47–48.

Woolcock, P. (1993). Ruse's darwinian meta-ethics: A critique. *Biology and Philosophy* 8: 423–439.

Wurtman, R. J. (1983). Food consumption, neurotransmitter synthesis, and human behavior. *Experientia* (Suppl.) 44: 356–369.

Young, S. N. (1993). The use of diet and dietary components in the study of factors controlling affect in humans: A review. *Journal of Psychiatry and Neuroscience* 18: 235–244.

Zahavi, A. (1995). Altruism as a handicap: The limitations of kin selection and reciprocity. *Journal of Avian Biology* 26: 1–3.

Zeisel, S. H. (1986). Dietary influences on neurotransmission. *Advances in Pediatrics* 33: 23–47.

Index

abortion, 113, 114, 115
Abraham, J., 103, 144
A-constraint, 43, 44
adaptation, 21, 26–28, 31, 51–53, 93, 106–109, 122, 134, 150, 160–163; analysis, 108; behavioral, 157; biological, 105; constraints on, 42, 43, 45, 47, 53; continuing, 91, 95; evolutionary, 127; explanation, 32, 35, 106; genetic, 26, 36; physiological, 26; relative, 19, 48; temperature, 63
adaptationism, 3, 25–37, 49, 53, 162, 164
adaptive capacity, 20, 21
adaptive explanation, 106, 107
adaptive peak, 46
adaptive problem, 106
adoption, 84; study, 153
A-fitness, 20
agent, contributive, 31, 35, 36, 37
aggression, 152, 154–159
agriculture, 101, 140, 141
allele, 17, 58, 60, 100, 105; frequencies, 57, 59; sickle-cell, 151
allomeme, 100, 101; neutral, 101
altruism, 3–5, 66–67, 73–97, 102, 105, 119, 121, 125, 163; in animals, 82–85; apparent, 4, 83, 84, 126; and egoism, 74–75, 76–80; –empathy model, 81; evolutionary, 4, 89, 126; genuine, 3, 86, 87, 89, 90, 95, 96, 126; ordinary, 126; paradox of, 89–92, 92–96; psychological, 89, 93; reciprocal, 83–85, 90, 94–97, 126
Amundson, R., 43, 138, 139
ancestral environment, 106, 108, 109
anomaly, 88
anorexia, 108
antibiotics, 23, 128, 150, 160, 164
anxiety, 152, 160
Arnhart, L., 111, 112, 113
artifact, 99
assumption, 60
Axelrod, R., 85, 86

Beatty, J., 2, 16–17, 18, 21, 22, 41
behavior, 73–75, 80–89, 94–97, 119–122, 125–127; actual, 54; aggressive, 156–158; altruistic, 66, 74, 81–83, 102, 121, 125; antisocial, 156; cooperative, 96; destructive, 81; egoistic, 4, 74; foraging, 3 (maladaptive, 4); genetics, 152–159; helping, 81; human, 4 (altruistic, 102; social, 73); immoral, 111, 113, 127; impossible, 43–45; innateness of, 106; irrational, 87; maladaptive, 4, 41, 53, 54;

mental, abnormalities of, 152; nonoptimal, 43, 46–52, 54; optimal, 3, 4, 39–53, 162; possible, 43–47, 54; prosocial, 157; social, 73; unopportunistic, 87
behavioral phenotypes, 155
benchmark, 70
benefit, 67, 74–79, 82–88, 94, 144; and cost, 85, 86, 88, 94; first-order, 74, 75; primary, 75, 77, 81; secondary, 75, 77, 78, 81; second-order, 74, 78
bias, 47, 75, 83, 103, 104, 133, 138, 144, 145, 147, 153, 156, 157
biological theory, generality of, 8–11
biology, 1, 2; philosophy of, 7
biomedical research, 144, 145
biotechnology, 127
Björklund, M., 46, 49
Blackmore, S., 99, 102
Block, N., 154
Boyd, R., 91, 99
Bradie, M., 123, 131, 132, 133, 138
Brandon, R. N., 1, 18–21, 32, 35, 36, 55–57, 61–65, 106, 107
bulimia, 108
Buss, D. M., 105, 108, 109
Byerly, H., 19, 20, 21
byproduct theory, 91, 95

Cairns, R. B., 152, 157
calcium-channel antagonists, 104, 144
Campbell, D. T., 132, 138
Campbell, J. H., 119–121
Campbell, R., 119
cardiovascular disease, 151, 160
Cartwright, N., 1, 167
causal determination, 44
causal factor, 14, 30, 31, 88, 142–155, 158
causality, 57, 144, 154
causal processes, 4, 62
causal role, 65
causation, 13, 16, 17, 29, 30, 34, 43–46, 56–60, 72, 121, 154, 159
cause, 143; of evolutionary change, 15; natural selection as, 28; of pg-fitness, 23; proximate, 61, 63, 65; remote, 61, 63, 65; of survival, 14
ceteris paribus condition, 11, 15
charity, 83
cheater detection, 86, 106
cheating, 83–88, 106, 126

Chua, G., 104
cladogenesis, 70
claim: empirical, 5; normative, 5, 115, 127
cognition, 132–136, 157
cognitive ability, 132
cognitive capacity, 135, 139, 146
cognitive ethology, 45
common sense, 73, 76, 89, 96, 113–115, 118, 134, 136, 153, 154
community good, 114, 116
comparative mode, 100, 101
compatibilism, 44
competition, 120, 138; intergroup, 137; interspecific, 10; intrasexual, 108; intraspecific, 10, 30
competitive ability, 127
conceptual selection, 138
conditional probability, 61
consciousness, 99
consilience, 135; of inductions, 133, 136
constraint, 4, 39–54, 84, 85, 88, 96, 97, 101, 150, 151, 160, 162; on adaptation (A-constraint), 43, 44; controversies over, 49–54; environmental, 33; and free will, 42–45; generalized (G-constraint), 43, 44; genetic, 49; hard, 45–49; soft, 45–49
context, 7–9; dependence, 20, 27, 36, 93, 154; of equation, 4, 58–63; evolutionary, 92; of explication, 58, 59; genotypic, 57; independence, 67; of interpretation, 56, 60, 72
contingency, historical, 28
continuing adaptation theory, 91, 95
contributive agent, 31, 35–37
cooperation, 4, 85–89, 92, 96, 97, 137, 163; in hunting, 109
Cosmides, L., 85–89, 105, 107, 109
cost, and benefit, 85, 86, 88, 94
credit, 137
crime, 113
cultural selection, 99, 100–105, 163
cultural unit, 100
culture, 5, 81, 99–110, 120, 131, 133, 145, 146, 157, 163
currency, 39, 41, 42, 47, 48, 75, 78, 90, 93, 125

Daly, M., 154
Damuth, J., 66, 68
Darwin, C., 5, 8, 23, 30, 33, 134, 138, 150
Davis, J. O., 158

Index

Dawkins, R., 56, 100
decision rule, 85, 86, 88, 89, 155
defense mechanism, 150, 152
defensive response, 152
deforestation, 140–144
De Fries, J. C., 156
density-dependent fitness, 31
depression, 152, 158, 160, 164
design, 9, 12, 13, 19, 27, 84, 108, 109, 112, 117; defect, 150; fitness as, 8; optimal, 28; suboptimal, 150
Detsky, A. S., 104
development agency, 140, 141
dichotomy, 63, 65, 74, 81, 144, 157
diet, 47, 136, 144, 151, 155, 158, 159; therapy, 103, 104, 155
differential survival, 8, 22, 23, 28–31, 33, 36, 37, 61, 70
dimensionality, 57, 58
directional selection, 32, 61
disease, 149–160; infectious, 149
disposition, 17, 43, 45, 88, 89, 92, 95, 96, 114, 116, 121, 160
dispositional property, 17
diversification, 100
diversity, 2, 50, 103, 105, 142
domain-specificity, 86, 87, 109
drift, 26, 53, 71; genetic, 25, 33, 52, 53, 70
drug therapy, 103, 104
dummy variable, 20, 21
Durham, W. H., 99, 100, 102, 103, 105
dynamic energy budget, 1
dynamic sufficiency, 59

ecological balance, 128
ecological environment, 32
ecological theory, 141
ecology, 8, 14, 23, 29, 30, 32, 35–37, 105, 126, 127, 140, 141, 152, 159; evolutionary, 128; laws in, 1, 34; neglect of, 104, 144, 145, 149, 157
e-explanation, 12–23
effect hypothesis, 68
e-fitness, 3, 8, 12–16, 18, 21–23, 29, 35, 161, 162
EGE (generative evolution), 120, 121
egoism, 5, 73–97, 125, 126; ethical, 4, 5, 73–79; high-minded, 79; psychological, 4, 5, 73, 74, 78, 81, 82, 89, 93, 94, 96, 163
Eldredge, N., 70
eliminativist strategy, 90, 91

emergence, 68, 71
emergent features of species, 68
emergent group feature, 66
Emlen, J. M., 39
empathy, 81
empirical claim, 5, 7, 10, 19, 22, 48, 54, 79
empirical content, 9, 10, 18, 19, 21, 40, 41, 59, 72; of equation, 4
empirical system, 9, 41
Endler, J. A., 31, 32
energy intake, 39, 41, 47
engineering fitness (e-fitness), 3, 8, 12–16, 18, 21–23, 29, 35, 161, 162
environment, 3, 10, 13, 15, 16, 19, 21, 25–27; ecological, 32; external, 32; prenatal, 153; selective, 32, 106
environmental factor, 2, 8, 20, 27–37, 89, 152–160
environmental management, 127
environmental selective agents, 26
environmental variables, 21
epidemic, 128, 143
epigenesis, 63
epigenetic law, 64
epistemology, 3; evolutionary, 131–147; normative, 5, 139
equal environments assumption, 153
equation, 56–60; core, for selection, 20; of population genetics, 4, 29, 30, 56, 58–60, 72, 163
ethical egoism, 4, 5, 73, 76–82, 93, 96, 163
ethical principle, 112, 119, 125
ethics, 3–5, 74–81, 84, 93, 96, 140, 163, 164; evolutionary, 111–129; normative, 5, 122
etiological factor, 144, 145, 159
etiology, 5, 104, 144, 145, 156, 159, 164
evolution, 2, 25, 28, 39, 55, 67, 70, 99, 117, 151; cause of, 36; of cognitive mechanisms program (EEM), 131–147; generative, 121; as genetic change, 4; of theories program (EET), 131–147
evolutionary biology, 62, 63, 164; and altruism, 4, 73–97, 119, 126, 163; and culture, 5, 105, 105–110; e-fitness as placeholder concept in, 13; and empirical claims, 5, 7; and ethical principles, 112–129, 163, 164; explanatory relevance of, 125–127; extensions of, 3; and fitness, 11, 22, 125 (as reproductive success, 19, 161); general concepts of, 161; general theories in, 22; historical changes in

genetic factors in, 2; and human knowledge, 131–147; and many-place predicate, 162; and medicine, 164; models of population genetics in, 4, 10; modern (and laws, 2, 29, 65); and natural history, 3, 5, 8, 36; and natural selection, 3, 10, 15, 31, 119, 162 (as law, 2, 7, 8, 10); in normative setting, 127, 164; and population genetics, 8, 95, 105; and psychopathology, 149; role of environment in, 32; and selfishness, 4, 73–97, 163; and systematics, 143; and tautology problem, 8, 10, 22, 161; theorizing in, 2; two stages of explanation in, 23
evolutionary contingency thesis, 2
evolutionary ecology, 128
evolutionary epistemology, 139, 140, 144, 146, 147, 164
evolutionary ethics, 111–129
evolutionary explanation; and altruism, 121; concrete, 61; and culture, 108, 109, 163; and fitness, 8, 12, 18, 20, 23; generality of, 8; of knowledge, 132, 134, 146; of morality, 112, 123, 128; and natural history, 35, 69, 125; and naturalistic fallacy, 129; and natural selection, 162; of origin of norms in ethics, 5; and psychopathology, 149, 159, 160
evolutionary foundation for ethics, 118
evolutionary progress, 121
evolutionary theory, 2, 3, 7–11, 22, 29, 32, 55–71, 73–96, 99–110, 111–119, 134–139, 150, 160
Ewald, P. W., 150
exaptation, 109
existential claim, 15, 162
existential component, 14, 16
expected survival, 16
explanation, 18, 76–97, 125–129; of adaptation, 32, 35, 48, 106, 107; of altruism and egoism, 4, 76, 80; of aspects of culture, 5; biological, 119; causal, 29, 36, 62, 69, 122; of changes in population, 8; circular, 12; of differential survival, 30; e- (ecological), 12–23; evolutionary, see evolutionary explanation; of features of disease, 150; of fitness, 12; functional, 50; genic, 63; of heavy-metal tolerance, 107; of human behavior, 105; of industrial melanism, 13, 14; vs. justification, 116; of mate preference, 108; "memetic," 102; of morality, 119, 122; of natural selection, 105; of nonmolecular traits, 53; pg- (population genetics), 12–23; of polymorphism, 60; of psychological processes, 105; of psychopathology, 153; of reproductive success, 3; scientific, 19; selective, 138; of sickle-cell anemia, 18, 34, 101; substantive, 3
exponential growth, 9
external environment, 32
externalist, 137

Fairhead, J., 140, 141, 142, 145
fallacy of changing subject, 124
fatty acid, 104
fertility selection, 64
fever, 150
F-fitness, 21, 22
field equation, 2
first-order benefit, 74, 75
fitness, 3, 7–23, 30–36; biological, 100, 101, 118; cultural, 100, 101; differences, heritable, 11, 15; inclusive, 83, 90, 92, 101, 137; maximization of, 42, 125 (and altruism; enhancement of, 88; low, 127; species-level, 68); population genetics (pg-fitness), 12, 14–23, 30–33; value, 66
food: allergy, 144; processing, 104, 159; see also diet
foraging, 30, 39–54, 68, 84; behavior, 4, 39, 45–51; genetics of, 49; optimal, theory (OFT), 3, 39–54
foundation, 3, 5, 87, 123; evolutionary, for ethics, 111, 112, 118, 120, 121
founder principle, 138
Fox Keller, E., 68
Frank, R. H., 87
fraud, 104
free-riding, 77
free will, 42–45, 44
frequency-dependent fitness, 31
frequency-dependent selection, 28, 60, 67, 107
friendship, 79, 97, 112

game theory, 85, 87, 126
gametic selection, 58
gang warfare, 108
Garrett, L., 104, 128, 145, 157

Index

Gauthier, D., 77, 78
G-constraint, 43, 44
gene, 63; -culture relationship, 100; flow, 106; frequencies, 4, 59, 163; sickle-cell, 101
general adaptedness, 20
generalist approach, 145–147
generality, 1–5, 7, 14–22, 93, 154, 161–163; and altruism and egoism, 75–77, 80–81; of biological theories, 8–11; and cultural evolution, 100; and ethics, 96–97, 124–125; misguided, 74; and optimality, 49–52; and scientific progress, 137–143; and universality and testability, 39–45
generalization, 61
generative evolution (EGE), 120, 121
genetic adaptation, 26, 27, 36
genetically modified organism, 127, 129
genetic determination, 153, 154, 157
genetic drift, 25, 33, 52, 53, 70
genetic engineering, 120
genetic factor, 2, 5, 27, 89, 144, 151–160, 164
genic selectionism, 4, 55–72, 163
genotoken, 64
genotype, 18–21, 28, 31–33, 36, 55–65, 71, 107, 138, 151; feature, 62–65; frequency, 57, 59
geographic isolation, 71, 138
Gewirth, A., 114, 116, 117, 118
Gilbert, P., 152
Godfrey-Smith, P., 56–60
Goodwin, B. C., 2
Gould, S. J., 25, 26, 27, 28, 31, 36, 37, 49, 107, 109
Grantham, T. A., 68, 69, 70, 71
Gray, R. D., 50, 63
group selection, 4, 65–72, 83, 84, 90, 92, 93, 126, 162, 163; cultural, 91, 95
growth: exponential, 9; logistic, 9

Haccou, P., 40
Hahlweg, K., 138
hard-wiring, 105
Harris, C. R., 109
hedonism, 93; psychological, 90
Heisler, I. L., 66, 68
heritability, 19, 27, 32, 55, 66, 67, 106–108, 154–156
heterosis, 17
heterozygote, 101; inferiority, 57–59; superiority, 17, 29, 34, 60
heuristics, 25, 88
Hinde, R. A., 109
historical contingency, 28, 52
Holmes, S., 80, 81, 87
homozygote, 57
honesty, 87
Hughes, W., 114, 124
Hull, D. L., 56, 136–139, 142–144
human behavior, 73, 75, 82, 85, 86, 88, 95–97, 105, 109, 163
human nature, 75, 81, 96, 105, 111–113, 117, 120, 122, 127
Hume, D., 80, 81, 111, 122
hunting, cooperation in, 109
hypothesis testing, 48

ideal system, 9, 41
imitation, 102
immoral behavior, 113, 116, 118, 121, 125–128
incest, 119; taboo, 101
inclusive fitness, 83, 90, 92, 101, 137
individual(s): and altruism, 66, 126; and behavior genetics, 153, 154, 159; causation of particular features of, 154; and extinction of species, 69, 71; fallacious inferences from populations to, 156; fitness of, 70, 96; and game theory, 85; invariant constraints across, 46; mate value of, 108; and natural selection, 55, 65; and OFT, 41; and phenotypes, 31; and predation, 31, 35; and productivity (reproductive output), 66; reproductive success of (pg-fitness), 11, 15, 19, 90; and speciation, 69
individual differences in a population, 107
individual selection vs. group selection, 65–68, 67, 71, 83, 92, 163
industrial melanism, 13, 14, 23, 35, 55
inference principle, 113–115
inference rule, 114–118
infinite regress, 114, 124, 128
inheritance, 18, 100
initial condition, 2
innate fear, 109
innovation, 100; cognitive, 107
instantiation, 9, 13, 19, 21, 114
intelligence, 120, 121, 126, 156; universal general, 106
interaction, 72, 136, 142

interactive mode, 100, 101
interactor, 56, 65, 72, 136–138
interdisciplinary link, 145
internalist, 137

Joseph, J., 153
justification, 114–129, 132–135, 146, 164
just-so stories, 27, 28, 107

Kantorovich, A., 138, 142
kin selection, 67, 83, 84, 90, 91, 92, 93, 95, 96, 126
Kitcher, P., 10, 14, 92, 112, 123
knowledge, 131–147; indigenous, 141
Kooijman, S. A. L. M., 1, 172
Kornblith, H., 132, 133
Krebs, J. R., 41, 42, 51

lactose absorption, 101
law, 2, 3, 7, 14, 60; epigenetic, 64; of evolution, 7, 10, 11; general, 1, 2, 14, 29, 75; Mendel's, 41; of nature, 7, 9, 42, 56, 72, 133, 137 (general, 36; universal, 1); phenomenological, 1; of population growth, 10; of transformation, 64
Lawton, J. H., 1
Leach, M., 140–142, 145
learning, 45, 46, 138, 157
Lemos, N. M., 79
Lennox, J. G., 33
Levins, R., 139, 140
Lewontin, R. C., 25–28, 31, 36, 37, 43, 47, 49, 52, 56–60, 64, 107
lineage, 70, 120, 136
locus, 57, 58
logic, 86, 87, 114–116, 121
logical reasoning, 87
logistic growth, 9
Lumsden, C. J., 99, 101, 112

macroevolution, 71; effect, 68, 69
maladaptation, 47, 53, 162
maladaptive beliefs, 134
maladaptive body-image disorders, 108
maladaptive features, 25, 53, 151
maladaptive traits, 28, 135
malaria, and sickle-cell anemia, 17, 18, 34, 101, 151, 160
many-place predicate, 26, 27, 40, 162, 163
Marks, I. M., 152
Martinsen, D., 118

mate value, 108
mathematical model, 10
mating, 13, 32, 57–59, 68, 108, 112
Maynard Smith, J., 41, 51, 52
McArthur, R. H., 39
medical research, 103, 104, 144, 147
medicine, 104, 128, 144–147, 149–160, 164; Darwinian, 5, 150–152
meme, 100, 102
memetics, 102
metamodel, 20
metaphysical explanation, 76
metatheory, 14, 21, 22, 142
methodological criterion, 9, 40, 76, 124–125, 140, 142
methodology, 104, 133–139, 143–145, 147
Michod, R., 19, 20, 21, 28
migration, 8–12, 18, 19, 23, 34, 36, 58, 128
Mills, S., 16–18, 21
model, 40–54, 56–60, 65; altruism–empathy, 81; assumption, 50, 51; -building, 26; context-dependent, 20; empathy–altruism, 81; genotypic, 56, 57; mathematical, 10; OFT, 45; of population genetics, 19, 30, 33, 34, 36 (abstract, 30; elementary, 30; equations in, 60; in evolutionary biology, 4; integrative, 64; for organismic selection, 69); prediction, 26, 49–51; standard social science (SSSM), 105, 106
modern synthesis, 65
modus ponens, 114–116
monomarital principle, 101
moral agency, 119
moral behavior, 111, 113, 117, 121, 125, 126
moral belief, 123
morality, 76–79, 94, 111–113, 116–129
moral principle, 118, 119, 123, 127
moral rightness, 79, 118
moral rule, 121, 123
motivation, 74–81, 87, 91, 93, 94, 97
motive, 103
mutation, 8, 11, 12, 25–28, 53, 55, 58, 70, 100, 138, 158
mutualism, 83–85, 96, 97, 126, 127

Nagel, T., 76
natural history, passim
naturalistic fallacy, 5, 111–124, 128, 129, 164
natural selection, passim; as cause, 28;

Index

diversifying, 99; principle of (PNS), 2, 3, 7, 10, 11, 15, 19, 21, 35, 36, 161; and psychopathology, 152–156
Nazi Germany, 121
necessary condition, 116, 118
Nesse, R. M., 150–152, 159
neurotransmission, 152, 157, 158
niche, 107
nonsteroid antiinflammatory drug, 103
norm, 111, 114; epistemological, 5, 163, 164; ethical, 5, 163, 164; of knowledge, 133; regulatory, 152; of science, 133; of selection, 142
normative claim, 5, 115, 127

one-place predicate, 26, 162
ontogeny, 138
ontology, 62, 65
optimal design, 28
optimal foraging theory, 39–54
optimality, 3, 4, 39–54, 162
organismic selection, 68–71
organization, level of, 65, 145
O'Rourke, F., 104
Orzack, S. H., 52, 53
Oster, G. F., 49
overconsumption, 151

panselectionism, 50
parameter value, 57
Parker, G. A., 51, 52
Pashler, H. E., 109
patch, 41
pathogen, 128, 149, 150, 158, 160, 164
payoff, 92
pg-explanation, 12–23
pg-fitness, 12, 14–23, 30–33
pharmaceutical industry, 103, 104, 144, 147
Phelps, J. A., 158
phenotoken, 62, 64
phenotype, 16–19, 26–36, 51, 53, 55, 57, 71, 107, 138, 142, 151–155; feature, 19, 29–31, 34, 35, 60–65, 151, 153 (contributive, 31; selective, 31)
phenotypic plasticity, 27, 89, 107
phenylketonuria (PKU), 155
phobia, 109
phylogeny, 95, 106, 107, 126, 138
physics, 1, 42
physiological adaptation, 26, 27, 36
Pianka, E. R., 39

placeholder, 8, 26, 36, 96, 161, 162; concept, 3, 13, 27, 29, 40, 75, 92, 94
pleiotropy, 25, 26, 35, 51, 84
Plomin, R., 156
polyandry, 101
polygyny, 101
polymorphism, 17, 60; genetic, 89, 107
Popper, K. R., 138
population, 8–12, 15–21, 29–36, 63–71, 151–160; change, 12, 15, 19, 20, 34, 36, 55; density, 26, 30, 31, 33; dynamics, 10; genetics, 8, 10–12, 19, 22, 23, 29, 30, 56–60, 68, 72, 84, 95 (abstract, 30; elementary, 30; equations in, 60; in evolutionary biology, 4; integrative, 64; model of, 19, 33 36; for organismic selection, 69); growth, 9, 10, 32, 33; phenotype, 55; polymorphic, 18; structure, 33, 106, 107; type, 3, 8, 11, 15, 19, 22, 23, 36, 161; variation in, 25
practical reason, 76
precision, 140
predation, 13, 14, 31, 32, 35, 55
predator, 13, 31, 55; detection, 68
predicate: many-place, 26; one-place, 26
prediction, 128; failed, 48, 49; model, 26, 49, 50, 51
prenatal environment, 153
prey, 41, 46, 53; selection, 39
Price, J., 151
primary benefit, 77, 81
prisoner's dilemma, 85, 87, 91, 95
probability, 12, 16, 31, 61, 62, 109, 151, 158; distribution, 22
process structuralism, 2
process, causal, 62
progress, 134, 137–139, 143, 147; evolutionary, 120, 121; moral, 120; in science, 137, 139, 142, 146, 163
propensity: concept of fitness, 8, 16–18, 21–23; interpretation, 14, 19, 21 (of fitness, 16–18)
proximate cause, 61, 62, 63, 65
psychiatric disorder, 144, 151, 152, 157–160, 164
psychiatry, 144, 149, 153, 164; biological, 152; evolutionary, 5
psychological egoism, 4, 5, 73, 74, 78, 81, 82, 89, 93, 94, 96, 163
psychology, 81, 84, 94, 144; developmental, 82; evolutionary, 74, 85–89, 105–109;

humanistic, 82; moral, 80; social, 82
psychopath, 113
psychopathology, 149–152, 155–160; and aggression, 157–159; and natural selection, 152–156
psychosocial factor, 144, 157
psychotherapy, 81, 82

rationality, 78, 79, 91, 95, 107, 117, 134, 135, 139, 146
realism, 134, 135, 140, 145, 147
reconciliationist strategy, 90, 91
reconstruction, 10–14, 19, 21, 32, 36, 51–54, 115
reductionism, 80, 145
reference, 80
reification, 109
remote cause, 61–65
replication, 56, 63, 64, 72, 99, 136, 142
replicator, 65, 72, 102, 136, 137, 138; selection, 56, 58
representation, 56–60, 72
reproduction, sexual, 29
reproductive success, 3, 8, 15, 18, 19, 25–29, 32–39, 61–65, 90, 97, 161; long-term, 22
reproductive survival, fitness as, 11–12
Rescher, N., 78, 79
resource, 10, 101, 108, 151
r-fitness, 21
rheumatoid arthritis, 103, 104
Richards, R. J., 113–119
Richardson, R. C., 21, 74, 107
Richerson, P. J., 91, 99
Rottschaefer, W. A., 118, 119, 125
rules of transformation, 57
Ruse, M., 121–123, 133–135

salience, 46, 145, 158–160, 163
schizoid personality, 151
schizophrenia, 152–160, 164
Schlager, D., 152
Schoener, T. W., 39, 49, 50, 51
Scott, S., 77, 78
screening off, 56, 61–65
secondary benefit, 75, 77, 78, 81
second-order benefit, 74, 78
selection: cultural, 99, 100–103, 163; directional, 32; fertility, 64; frequency-dependent, 28, 60; gametic, 58; group, 65–71; individual, 65–68, 71, 83, 92, 163; kin, 67; level of, 55–72; natural, passim; organismic, 68–71; prey, 39; process, 3–5, 28–31, 56–72, 132–139 (cultural, 104; natural, 104); replicator, 56, 58; species, 4, 65–72, 162, 163; theory, 139, 144; units of, 55–72; vehicle, 56; viability, 63, 64
selection for, 35, 108
selection of, 35, 108, 109
selectionism, 27; genic, 56–60
selective agent, 3, 26–37, 104, 105, 142, 161, 162
selective environment, 32, 33, 106
self, concept of, 75
self-awareness, 99
self-interest, 73–97
selfishness, 4, 73, 74, 78, 81, 82, 96, 125, 163; evolutionary, 126
semantic view, 10, 41; in philosophy, 9
Sesardic, N., 74, 89–97, 126
sexuality, 68
sexual reproduction, 29, 68
sickle-cell anemia, 17, 18, 34, 101, 151, 160
simplicity, 76, 80
skepticism, 134, 135
Sloep, P. B., 10, 42
Smith, A., 80, 81
Sober, E., 2, 17–18, 22, 26–28, 34, 35, 52–57, 61–71, 83, 90–93, 108, 122, 125, 133
social behavior, 84, 112, 121, 157
sociobiology, 112, 124
sociology of science, 139
source law, 34
specialist approach, 145–147
speciation, 68–71, 138
species, 8, 20; selection, 4, 65–72, 162, 163; sorting, 68, 71
spiritualism, 122
standard social science model (SSSM), 105, 106
Stanley, S., 70, 71
state space, 57, 59
Stelfox, H. T., 104, 144
Stephens, D. W., 41, 42, 51
Stevens, A., 151
Stidd, B. M., 71
Stoff, D. M., 152, 157
strategy set, 39, 40, 43, 44, 47, 48, 53, 54, 162
struggle for existence, 30, 33

subjectivism, 122
succession theory, 140, 141, 142
sufficiency, dynamic, 59
sufficient condition, 116, 118
supervenience, 12–17, 21–23
survival, 7–23, 28–33, 55, 61, 97, 118, 133–135, 151, 161; actual, 12; differential, 8, 22, 23, 28–31, 33, 36, 37, 61, 70; expected, 16; of the fittest, 8, 120; reproductive, 11–12, 87, 89, 95 (design for, 8; differential, 55); value, 132, 133, 135
sustainability, 140
system: empirical, 9; ideal, 9
systematics, 143, 144

tautology, 3, 5, 8–15, 19, 20, 22, 23, 78, 80, 81, 88, 93, 134, 161, 162
Taylor, P. W., 78, 79
Tennant, N., 138, 142
testability, 40–42, 50, 51, 52, 54
theoretical hypothesis, 9, 10, 41, 42
theory, 1, 9, 42, 65; byproduct, 91, 95; continuing adaptation, 91, 95; core, general, 54; ecological, 141; evolutionary, 29; general, 4, 13, 35, 39, 44, 45, 50, 52, 72; optimality, 53; selection, 139, 144–146; succession, 140; vestige, 91, 95
Thompson, P., 10, 41, 126, 133, 135, 136
tolerance, heavy-metal, 19, 35, 36, 107
Tooby, J., 85–89, 105, 107, 109
trade-off, 140, 142, 150–152
transformation, rule of, 57
transmission, 101, 104; unit of, 100
truth, 123, 124, 132–137, 140, 147
twin study, 152

types, in populations, persistence of, 8, 11, 15, 19, 22, 161

underdetermination, 135, 136, 145
unification, 19, 20
universality, 1, 3, 7, 40–42, 54, 60, 96, 105, 107
universal quantifier, 40

validity, 42
value, 120, 121, 124; cultural, 118
variability, 60, 89, 105, 107, 108, 113
variation, sources of, 100; environmental, 154; genetic, 154
Vbra, E. S., 68, 71, 109
vehicle, 58, 65, 72; vehicle selection, 56
vestige theory, 91, 95
viability selection, 63, 64
virulence, 150

Wade, M. J., 66, 71
Wallach, L., 81, 82, 126
Wallach, M. A., 81, 82, 126
war, 102, 113; crime, 113
Wason selection task, 86
Weber, M., 1
Williams, G. C., 150, 151, 152, 159
Williams, P. C., 114, 118
Williams, T. D., 84
Wilson, B. E., 33
Wilson, D. R., 151, 152
Wilson, D. S., 55, 62, 66–68, 83, 89, 90, 92, 107, 125
Wilson, E. O., 49, 99, 101, 112
worldview, 111

Zahavi, A., 83

About the Author

WIM J. VAN DER STEEN is Professor of Philosophy of Biology, Faculties of Biology and Philosophy and the Institute of Ethics, Vrije Universiteit, Amsterdam, and has published in experimental biology, the philosophy of biology, the philosophy of medicine, and ethics.